the RE-ORIGIN of SPECIES

Torill Kornfeldt is a Swedish science journalist with a background in biology. She has worked for Sweden's leading morning newspaper, *Dagens Nyheter*, and for the Sveriges Radio public broadcaster, where she created the successful radio show *Tekniksafari*. Her focus is on how emerging bioengineering and technology will shape our future. *The Re-Origin of Species* is her first book.

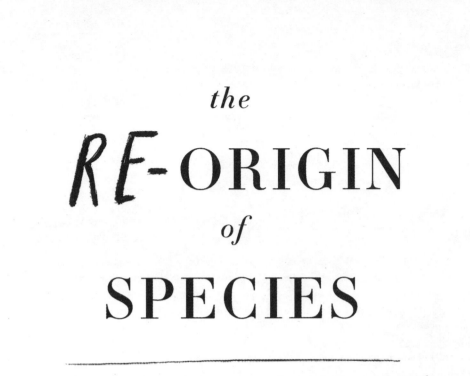

the

RE-ORIGIN

of

SPECIES

A SECOND CHANCE FOR EXTINCT ANIMALS

Torill Kornfeldt

TRANSLATED BY FIONA GRAHAM

SCRIBE

Melbourne • London

Scribe Publications
2 John Street, Clerkenwell, London, WC1N 2ES, United Kingdom
18–20 Edward St, Brunswick, Victoria 3056, Australia

Originally published as *Mammutens återkomst* in Swedish by Fri Tanke Förlag,
Sweden in 2016

First published in English by Scribe in 2018
Published by agreement with the Kontext Agency

Typset in Garamond Premier Pro by J&M Typesetting P/L
Printed and bound in the UK by CPI Group (UK) Ltd, Croydon CR0 4YY

Scribe Publications is committed to the sustainable use of natural resources and the
use of paper products made responsibly from those resources.

The cost of this translation was defrayed by a subsidy from the Swedish Arts
Council, gratefully acknowledged.

9781911617228 (UK edition)
9781947534360 (US edition)
9781925713060 (ANZ edition)
9781925693003 (e-book)

CiP data records for this title are available from the National Library of Australia
and the British Library.

scribepublications.co.uk
scribepublications.com
scribepublications.com.au

for
Tobias, Torgny, and Ruth-Aimée

TRANSLATOR'S NOTE

The passages quoting those scientists who were interviewed in English are not verbatim transcriptions. In many cases, they are a compressed or summarised representation of what the interviewees said, based sometimes on sound recordings and sometimes on written notes. It was not possible to have access to all of this material.

CONTENTS

INTRODUCTION
A Whole New World

Greek mythology tells the tale of Prometheus, who defied the gods to bring humankind the knowledge of fire and how to use it. Prometheus was harshly punished by Zeus, yet the Greeks regarded fire as the source of all art and science. This narrative resembles the Biblical Fall; while the fruit of knowledge comes at a high price, it is a precondition for becoming fully human.

A couple of thousand years later, in 1818, Mary Shelley published her novel *Frankenstein; or, The Modern Prometheus*. The story showed what might happen if human pride and ambition overreached themselves in a bid to emulate God. At the time the book was written, scientists had just discovered that they could make a dead frog twitch by delivering jolts of electricity to the corpse. There were theories that a divine, life-giving force might have been discovered. Taking inspiration from Jewish legends about golems, Mary Shelley created a terrifying scenario in which a scientist applied that force without really comprehending or being able to control it. For the book is also the story of a scientist who refuses to accept responsibility for his work — a man who flees, leaving the newly awakened monster to his fate. It's clear that if Victor Frankenstein had had the courage to stay and take care of his creature, tragedy could have been averted.

One hundred and seventy-five years later, in 1993, the film *Jurassic Park* showed resurrected dinosaurs running amok because scientists had let their enthusiasm and curiosity get the better of them. The moral that revolutionary knowledge and godlike powers can cost us dear has been thoroughly rammed home. At the same time, the idea persists that we would no longer be human if we lacked that very drive.

It may sound silly to let yourself be influenced by ancient myths today, yet I believe these stories were the reason for my mixed feelings when I heard that scientists were working towards resurrecting extinct animals with the help of modern gene technology.

My first reaction was one of boundless enthusiasm. I felt like an excited ten-year-old at the thought of being able to see a real live mammoth, or a dinosaur, or any of the creatures that have died out in the course of history — at the idea of actually seeing them move and hearing the sounds they made. What does a mammoth smell like? Do dinosaurs bob their heads as they walk, like today's birds? Do aurochs low like cows?

Scientists are also trying to restore many less spectacular creatures that are at least equally fascinating. The Australian gastric-brooding frog is one example. The female frog would eat her eggs and let them develop from frogspawn into baby frogs inside her belly. Then she would regurgitate a litter of croaking froglets ready to meet the world. This genus died out in the 1980s, struck by a fungal disease that continues to threaten many other types of frog. The project to resurrect the gastric-brooding frog has been dubbed 'Lazarus', after the story of how Jesus brought a man back from the dead.

All the projects I describe in this book began with the

thought: 'Wow! We could actually do this. Of course we'll give it a try!' They are driven by the same enthusiasm and curiosity that makes a child learn the name of every dinosaur that ever walked the earth, or an explorer set sail for the distant horizon. It's easy to be swept along and to feel the same effervescent energy.

My second feeling was one of age-old unease. Is this really a good idea? What if it has unforeseen negative consequences? Would it mean unleashing forces that we would later be unable to control? This unease is not born of mythology alone. There's no shortage of examples of well-intentioned people who have wreaked havoc on the natural world.

One example — absurd from today's perspective, but a good illustration — is afforded by Eugene Schieffelin, an American who began releasing European birds in New York in 1890. His aim was to make sure that every single type of bird mentioned in the works of Shakespeare was represented in the United States. Schieffelin was a member of a highly respected scientific association, and his project enjoyed widespread support. It was associated with the acclimatisation movement, which actively dispersed species from continent to continent.

While most of the species died out within just a few years, the hundred starlings released in Central Park multiplied rapidly. They displaced large numbers of native birds as they spread out over the continent. Today, the United States has about 200 million European starlings, creating problems both for the ecosystem and for farming. And all this sprang from the worthiest of aims — biological and cultural enrichment.

Each day as a science journalist brings me examples of how scientific curiosity and zeal are improving life for almost all human beings. These range from the technology we use to the medicines we take, the food we eat and the clothes we wear. I am genuinely convinced that the world is getting better all the time, and that the main reason for this is new research. And yet — for all my optimism and confidence in the future — that uneasy feeling in my stomach persists.

Gene technology and biotechnology are developing at the same rate today as information technology did in the 1990s, possibly even faster. This means scientists can already do things considered impossible just a few years ago. It also means they will very soon be able to achieve things that seem impossible today. Recreating the mammoth may be one example.

New methods for restructuring genetic material in everything from bacteria to human beings have created a whole new world of opportunities, but also of fears. This potential seems all the more terrifying because of its novelty. Just as with the advent of computers, we lack a context for this development that could help us understand it and predict where it will lead.

I believe that genetics and biotech are going to transform our society in just as fundamental a way as digital technology has done. I am also convinced that most of that change will be positive. At the same time, major problems are bound to arise. I don't think we will make any progress unless we take this fear seriously, examine it, and analyse the cases in which it is relevant. This means looking at the practical aspects: how can we avoid making the same kind of mistakes as Victor Frankenstein or Eugene Schieffelin? But it also means examining the philosophical side: how will the capacity to manipulate life affect us as human beings, our culture and our society?

The third thought that came to mind when I first heard about these projects was that the desire to bring extinct creatures back to life sprang from nostalgia, a yearning to return to a lost world. I have met old men who seem to be dreaming of immortality. Four of the keenest researchers in this field are in their 60s. I have also met Ben, not yet 30, who has resolved to devote the rest of his career to resurrecting an extinct species of pigeon. All of them have a deep-seated feeling that the world and humanity have lost something important, and that there is a chance we could recover it. Exactly what it is we have lost, and when it happened, are questions to which they all have different answers.

Those three emotions — enthusiasm, fear, and nostalgia — have accompanied me throughout my work on this book. However, I have also realised that there is more, much more, to tell about the scientists who are determined to try to bring creatures back from the dead. And there is another aspect to their efforts that is arguably even more significant.

All the scientists I have spoken with seek to make the world a richer, wilder, and better place. They are convinced that reviving extinct fauna can contribute to such a future. Henri, who aims to breed an aurochs; George, who is trying to piece together a mammoth; William, who wants to see majestic American chestnut trees again; and so on. All of them are aiming to create a whole species that can be returned to nature, not merely a single individual.

The sole exception is Jack, who is trying to recreate a dinosaur. That experiment is different from all the rest, so if the only reason you picked up this book was to read about the chances of there being a real-life Jurassic Park, I suggest you go straight to Chapter 13. I hope you will find it so fascinating that you will come back to the beginning. And if you want to know more about the various projects, there are sources and notes at the back of this book and on the associated website.

It remains to be seen how resurrecting a species would work in practice. Essentially, all the projects I refer to in the book depend on at least one major scientific breakthrough in the future if they are to succeed. However, such breakthroughs are now coming so thick and fast that it is hard to see this as much of an obstacle.

What really fascinates me in the idea of reviving extinct creatures is that the mere thought of it expands my horizons, opening up dazzling new possibilities. Yet there is a fundamental question we need to ask ourselves collectively, and that is how far human beings should go in controlling nature. Now that we are on the threshold of being able to recreate lost creatures, reconstruct wild species, create entirely new forms of life that would never have come into being unaided — what do we do with that knowledge?

Is it a good idea to resurrect lost animal species? I shall do my best to explain how this aim might be achieved, and then you may answer the question yourself.

CHAPTER 1

Summer in Siberia

The only way to reach Chersky in eastern Siberia is to fly there in a small, battered prop plane. There are twice-weekly flights from the new airport in Yakutsk, the coldest city on Earth. In winter, the temperature here can fall to minus 50 degrees Celsius, but today, in mid-July, it's oppressively hot.

We are sitting in a small bus, waiting to board: 13 adults, two children, and a tiny dog with tufty paws and ears. A man holds an orchid in a plant pot; a woman grasps what appears to be a Christmas decoration the same size as herself, wrapped in a black plastic sack; another woman has a set of curtain rails. I am the only non-Russian-speaker, the only one not heading home after a shopping expedition to the metropolis of Yakutsk.

The plane looks as though it might fall apart at any moment, and a mechanic in dungarees wanders around, poking about under the inspection hatches with a screwdriver. One of the pilots goes and fumbles with the propellers to check whether they will turn. I sit in the bus, growing increasingly nervous. Should I decide not to board the plane after all? But what else can I do? After all, this is the only way to get to Chersky, and no one else seems to be particularly concerned about the safety of the flight. Finally, I climb the rickety steps along with everyone else.

No one pays any attention to the seat numbers printed on the tickets. The two female cabin attendants instruct us to sit right at the front. They speak no English, but point and gesture. The ramshackle seats are so worn-out that the backrest won't stay in place, so we passengers spend the whole journey semi-recumbent. The promised lifejacket supposedly stowed under the seat is conspicuous by its absence. The cabin staff walk along the narrow aisle distributing sick-bags and coffee, while the little dog scampers around between the seats. The plane shakes and rattles menacingly, but once airborne it flies smoothly, setting a course almost due east. Nonetheless, my pulse rate is higher than usual for the five hours of the flight.

'That plane hasn't crashed in 50 years,' says Nikita Zimov once I'm back on terra firma. 'So why should it crash this time?'

We are sitting in the spacious, round common room that is the heart of the research station I have come to Chersky to visit. It was Nikita's father, Sergey, who set up the station in the 1980s. It is a few kilometres outside the town — which itself is about as far away from anywhere else as you can possibly get.

This is the Siberian back country; the town lies to the north and slightly to the east of Japan, but not quite as far east as the Kamchatka Peninsula. To reach the northern, Arctic Ocean coast takes a few days by boat along the broad River Kolyma. There are no roads to Chersky; the only way to get here is by plane or boat. Prisoners were sent here in Soviet times, and the Russian Gold Rush came here in Chersky's boomtown days. Now about a third of the houses are abandoned, and the population has shrunk to barely 3,000. I'm told there were two swimming baths for a while in the '80s, but now they're gone, just like the restaurants.

Setting aside the decrepit buildings in town, this is a very beautiful place: a broad, flat landscape full of meandering rivers and shallow lakes. Sallow and larch woods cover the land beyond the river floodplains. Succulent green tufts of grass grow in stretches of shallow mud. The river bends enclose long shorelines, and there are bushy dwarf birches on the hillsides, where the soil is dryer. Now, in July, rosebay willowherb and tansy, bright-pink carnations and blue spike speedwells are in bloom everywhere.

'I've heard you Swedes can hold your liquor,' says Nikita, handing me a shot of vodka on my first evening. Everyone drinks vodka at dinner. Sergey has at least one glass over lunch.

It would be easy to resort to stereotypes and clichés in attempting to describe Sergey Zimov. A Russian scientist living an isolated life in the remote wilds of Siberia, he has long grey hair and a beard that is almost equally long and grey. He flits around the station in a T-shirt, a beret on his head and a cigarette in the corner of his mouth. His wife, Galina, deals with most of the paperwork.

Sergey has a whole set of opinions on what is fitting for each sex. But he is by no means the only one; throughout my time here, for example, I can't get into a boat or climb out of one without being offered a helping hand. Sergey is clearly proud of having a son, Nikita, to take over the station. He talks rather less about his daughter, a novelist living in St Petersburg. But there's nothing wrong with women scientists — some of the best scientists to visit the station have been women, Sergey tells me on the first evening.

He began his research out here at a time when the Soviet Union was doing its utmost to spread resources and influence in northern Siberia. That was an attempt to cement Russian influence over the country as a whole; Russian is not the mother tongue of the locals in this area, who have a written language of their own. 'Ethnic Russians' were sent here in an attempt to hold the country together, and quantities of research stations, mines, and other projects were set up, while air traffic expanded at the same time.

'This was a good place to be. I had plenty of freedom, and I was a long way away from any Communist propaganda,' he says over an evening meal of elk burgers.

The food at the station is excellent, provided you like elk meat (moose to North Americans). In the evenings, when we drink beer and play cards, everyone chews dried, salted squid. It's tasty, if a little tough.

When the Soviet Union fell apart, support for the station dried up. Sergey was ordered to pack up his family, leave the station, and return to the University of Novosibirsk. He refused. Instead, he decided to stay and set up Russia's first private research station, together with his family.

It was hard in the beginning. Nikita recalls the 1990s, when he was a teenager, as a grey time. The family could barely afford food sometimes. The situation is different now, with 50-odd international scientists, mainly Americans, coming here each year to study the natural environment and the permafrost. I am one of about 15 visitors, including some German scientists and a group of students from the US who play their guitars in the evenings.

'The hero of *Forrest Gump* becomes a successful shrimp fisherman by chance, just because a storm has destroyed all the other boats. It was like that for us — there are still very few research stations in the north with our capacity,' says Nikita.

The reason I have come all the way here is to look at mammoths, or at least at what remains of the ecosystem they lived in. About ten different species of mammoths have evolved and died out over the last five million years, the last of which was the woolly mammoth. That's the one most of us think of when we hear the word: a massive creature with a sloping back, covered in thick, curly fur and sporting huge, curved tusks. It evolved from its precursors about 400,000 years ago, somewhere in east Asia.

Mammoths inhabited a vast area stretching from today's Spain, Italy, and southern Sweden, through the whole of Siberia and large tracts of China, into Alaska and North America. In all probability, they lived in herds led by older females, like modern elephants. Humans encountered mammoths for the first time between 30,000 and 40,000 years ago, when we left Africa and made our way into the Middle East and Europe. By that time, the Neanderthals had already been living side-by-side with mammoths for a long time, hunting them, and sometimes using their bones as building material.

The last Ice Age began roughly 100,000 years ago. At the time when Scandinavia was covered by thick glaciers, a fertile steppe landscape was flourishing here in eastern Siberia. Winds and ocean currents made the region dry and windswept but kept it free of ice, so grass could grow during the warm summer months. The mammoths thrived here, together with woolly rhinoceroses, musk oxen, horses, and wolves. Nikita and Sergey have tried to calculate the number of animals that would have lived here some 40,000 years ago; according to their models, the wealth of fauna was almost as great as in the African savannah. When the first people arrived, about 27,000 years ago, there must have been almost infinite quantities of game for them to hunt.

About 10,000 years ago, the climate changed and the Ice Age ended. Siberia grew warmer, and, at about the same time, the mammoths vanished, too. Exactly why they disappeared remains unclear and is a subject of heated debate among scientists worldwide. Was it because of the warmer climate, or because the people multiplied and their hunting skills improved? It may have been a combination of the two, according to palaeogeneticist Beth Shapiro. Her research has shown that mammoths survived warm periods before the last Ice Age, though they dwindled in number, partly because the grasslands shrank and gave way to peat bogs and marshland. The same thing happened at the end of the last Ice Age, but Beth believes the presence of human hunters was the last nail in the mammoths' coffin. The question is a long way from being settled.

Neolithic cave painting of a mammoth, Rouffignac (in the Dordogne region of France).

Many other species disappeared at the same time as the mammoths, including the woolly rhinoceroses, and the broad, grassy steppes were replaced by today's wetlands and larch woods. Mammoths survived for much longer on a number of islands in the Arctic Ocean, especially Wrangel Island. The last one died about 4,000 years ago, a few hundred years after the great Egyptian pyramids at Giza were completed.

'At the time when there were most mammoths here, the ecosystem was so rich that it still provides local people with a livelihood,' says Sergey.

If you don't happen to run a research station, there are two ways to earn money in Chersky, he tells me. Either you fish the local char in the rivers, or you hunt for mammoth tusks. Many people have started searching for tusks in recent years. The prices that Chinese buyers are prepared to pay have risen rapidly. In Soviet times, such tusks were essentially worthless.

Some people in Chersky have bought diving equipment to go diving in the river, while others spend several months at a time in remote wilderness areas searching for tusks. An estimated 55 tonnes of mammoth ivory are shipped out of Siberia every year, nearly all ending up in China. The trade is legal, but a fairly high proportion of it is transacted on the black market so as to avoid tax and customs duties.

'Finding a tusk is the only way people here can afford an expensive item like a snow scooter,' says Sergey.

He has found umpteen tusks on his expeditions, he says, but, now they have a monetary value, they have become far rarer. Then he starts to tell me about the biggest tusk he's ever found.

'The base of it was this broad,' he says, measuring out a generous length with his hands — almost half a metre. 'And it was this long,' he continues, stretching out his arms.

I imagine this is Chersky's answer to fishermen's bragging tales, and that the tusk grows by a few centimetres for every journalist who gets to hear the story. However, it's true that mammoth tusks can be enormous. Both males and females bore tusks, but the females' were smaller and more slender. They spiralled as they grew, first growing straight out from the mammoth's head, then inwards again until they met, even crossing one another in some cases. The longest tusk ever found measures just over four metres.

In the corner of Sergey's living room stand two well-preserved tusks a metre long and two woolly rhino skulls.

'That's my insurance for if times get worse. The longest tusk is worth about 50,000 dollars,' he chuckles.

There are bits of mammoth all over the station, in fact. Chunks of mammoth tooth are used as paperweights here and there in the common room. Just like elephants, mammoths had huge teeth, and there were only four in their mouths at any one time: two in the upper jaw and two in the lower. A mammoth tooth can weigh nearly two kilograms.

There's a large cardboard box in the corridor between the toilets and the bedrooms, full of long mammoth bones lying higgledy-piggledy. Someone has written something indecipherable in Russian on the box with a felt pen, but judging by their shape, I guess the bones are femurs. I can't stop myself touching them each time I go past. Scientists hope to be able to recreate mammoths with the help of just such bones as these. The American palaeogeneticist Beth Shapiro, whom I mentioned earlier, is one of the world's main experts in extracting genetic material from Ice Age animal bones. She describes the process of fitting together the genes of creatures like mammoths as lengthy and difficult.

Imagine that an organism's hereditary material is a thick book, along the lines of *War and Peace*, *The Lord of the Rings*, or Shakespeare's collected works, and that it is present in every cell of a mammoth's body. But genetic material differs from books in that it needs to be constantly repaired in order to stay in one piece and remain legible, so there is an ongoing renovation process inside the cell. However, as soon as the mammoth dies, the long DNA molecules begin to decompose into ever-smaller fragments. Imagine that the glue holding the book together loosens, so the pages come apart. Then the paper starts to fall to bits, separating into individual sentences and words.

Imagine that you then pick up the resulting fragments and, as Shapiro puts it, you spread them out over a muddy field, in the rain, and have a horde of Ice Age animals trample them underfoot. Attempting to piece together Shakespeare's *Hamlet* out of the resultant mess is roughly what the people who sequence ancient DNA have to do.

Mammoth bones lie frozen in the permafrost, and they have lain buried in the earth for tens of thousands of years. They may be the remains of an animal that drowned in a mere and came to be embedded in frozen sediment. Although they are preserved by the cold, the mammoth's genetic material continues to disintegrate throughout that time. The ancient bones are ground down, enabling scientists to extract the remaining short sections of the DNA molecule — but now we come to the next problem. The muddy field where you are hunting for your book is strewn with other books. Scientists find quantities of DNA from bacteria, fungi, insects, and everything else that existed inside the bone at some point during the tens of thousands of years that it lay frozen in the tundra. In some cases, just 1 per cent of the DNA present comes from the mammoth itself.

Once you have identified all the minute fragments and established which ones actually come from the mammoth, you face the task of fitting them together in the right order. The only way to do this is to take a closely related species, such as the Asian elephant, as a kind of model. Each tiny piece of genetic material is compared with the model and inserted in the right place. Ultimately, you build up a kind of patchwork quilt in which each fragment of DNA overlaps with another fragment. This ultimately gives you an impression of how the mammoth's DNA would have looked.

This method has enabled scientists to assemble the mammoth's hereditary material step by step, and with increasing precision. The latest major study, which included scientists from the Swedish Museum of Natural History in Stockholm, involved sequencing the mammoth genome.

So scientists today know exactly what genes mammoths bear, and how they differ from those of the Asian elephant. It is possible to identify the genes that made mammoths different from elephants: those that gave them a thick coat, subcutaneous fat, smaller ears so that they could better retain heat, and so on. This knowledge raises hopes that it may be possible to recreate the mammoth.

Nikita and Sergey are not just researching the natural environment where mammoths once lived; rather, this is the habitat where new mammoths could live in future. The work of creating new shaggy-coated giants is already under way, and the cells that represent the first stage in that process are already growing in a Boston laboratory.

But before flying to the United States, I plan to visit the city of Yakutsk and its top tourist attraction.

CHAPTER 2

Who Wants to Build a Mammoth?

Yakutsk is not richly endowed with tourist attractions. Though one of the world's major diamond production centres and the capital of a region seven times the size of my homeland, Sweden, the city is home to barely 300,000 people. Its main square features an enormous statue of Lenin gazing towards the north and a large fountain that attracts teenagers at nightfall.

But the prime tourist trap, which lies some distance outside the city, is the 'Permafrost Kingdom'. Long caves lead down into the permanently frozen ground, their ceilings and walls covered in hoarfrost ten centimetres thick. All visitors are issued with a warm coat and winter footwear. Multicoloured lights illuminate ice statues, while classical music tinkles from the loudspeakers. You can buy vodka at a bar that is also made of ice. Besides vodka, it serves a traditional local speciality, *stroganina*, which is rather like sushi, consisting of chunks of frozen raw fish served with chopped onions, oil, and generous amounts of black pepper. The place offers charm, unashamed kitsch, and enough cold to freeze your hands.

Yet the thing that makes it truly fascinating is to be found in a little side room off the very first section of the caves. The guide speaks only a few words of English, but he unlocks a few doors and gestures to me to enter.

The walls are lined with blocks of ice that have been carved out and piled up. There's no music or colourful lighting here. A large grey mammoth head lies on a wooden pallet. Its trunk is missing, but the head is otherwise well-preserved. There are wrinkles around its eyes and a remaining tuft of dark brown hair on the crown. Its ears have been preserved, as have parts of its mouth.

Le Mammouth, by Paul Jamin, 1885.

There is a rather musty odour in here, a smell of dust and enclosed space, though it isn't strong enough to be disagreeable. The faint smell is the only indication that this mammoth died over 20,000 years ago and that the head in front of me is not fresh. The curling tusks diverge from the head; where the gap is widest, I can't reach from one tusk to the other. Further inside the room lies the body of a 30,000-year-old woolly rhino, but I have eyes only for the mammoth. I manoeuvre around it in the tight space so as to examine it from every possible angle, I touch its shiny tusks, and I stoop to get closer and scrutinise its wrinkly hide.

Now and then, researchers in Siberia find frozen cadavers like these. This particular one was discovered by a French mammoth hunter in 2002. It is the best-preserved head of a male mammoth ever found. But even better-preserved young mammoths have been found — whole bodies, looking almost as if they are only sleeping. The three best-known, Lyuba, Zhenya, and Dima, have toured museums worldwide. The thought that strikes me as I stand in this room — just as when I look at pictures of the mammoth calves — is that it can't be so very difficult to clone them.

The head next to me still looks so alive. Surely, I think, there just have to be some cells in there somewhere that, with a little scientific sleight of hand, could be brought to life.

These days, it has become almost commonplace for scientists to clone animals. The most usual method involves taking a cell nucleus from a full-grown animal and inserting it into an ovum or an embryonic cell. The cell nucleus, which contains the hereditary material, functions as the cell's control centre, directing operations. Full-grown cells are specialised; a skin cell, for instance, can't suddenly turn into a muscle cell. However, if a cell nucleus from a full-grown cell is inserted into an ovum, it can undergo a transformation, losing its specialised character. This enables it to guide the egg cell as it divides and grows, ultimately developing into a quite different animal. This is how Dolly the sheep was cloned more than 20 years ago. The method can be applied to animals that are similar or very closely related. Might it be possible to transfer a cell nucleus from a well-preserved mammoth corpse into a living elephant ovum?

There are several scientists who aim to do just that. They are searching the ice for bodies in increasingly good states of preservation. Sooner or later, they hope to find cell nuclei that are still alive, or, at any rate, sufficiently free of damage. As various lab experiments have shown, even if cells have lain frozen for years it is still possible to extract their cell nuclei and insert them into other cells when they thaw out.

The mammoth-cloning project that has attracted most attention is led by a South Korean, Hwang Woo-suk. In 2013, scientists found a mammoth carcass that was so well-preserved that it was oozing a blood-like liquid. The scientists concerned claim they will have cloned the first baby mammoth within just a few years.

However, this particular project needs to be taken with a very large pinch of salt. Hwang Woo-suk gained notoriety in the scientific world when, in 2004, he published a scientific article in which he claimed to have cloned 30 human embryos. This soon proved to be false, there being no such clones. Lengthy and complex legal proceedings ensued, tearing Hwang Woo-suk's scientific reputation to shreds. Yet he has made a comeback in recent years, and his research activities now include mammoth-hunting. Both the project as a whole and the latest find have attracted a lot of attention, but the researchers involved haven't yet published any scientific studies of attempts at cloning. Over the last few years there have been worrying signs of both legal and scientific problems. Many other geneticists worldwide have criticised the whole enterprise as being merely a way to attract funding and publicity, without there being any scientific basis.

There is another project, under the leadership of Japanese scientist Akira Iritani, an expert in reviving frozen cells. His exploits include cloning mice from cells that had lain frozen for 16 years. In 2011, he claimed that mammoths could be cloned 'within four to five years'. His team is also searching for the perfect carcass to harvest cells from. However, since no fluffy little baby elephant has yet been born at any zoo, this project seems to be a trickier enterprise than scientists initially imagined, and there have been no further updates.

So what makes it so hard to clone mammoths, when scientists have successfully sequenced their genome?

Imagine putting a piece of meat into the freezer, then taking it out, and repeating the process again and again. To begin with, it would have taken a long time for a mammoth's huge body to freeze, even under the most favourable circumstances. During that time, the cells began to break down and the flesh to decompose. Imagine, for example, that the body sank down into the sand at the bottom of a shallow lake; perhaps the mammoth got stuck while trying to wade across. The lake froze in winter, and the permafrost made it freeze from the bottom upwards, as well as from the top down. Next summer it thawed again, at least partially, and the process continued for several years until the body was finally covered with so much sediment from the lake bed and had sunk so far down that it remained frozen all year round. It could then lie frozen for 20,000 or 30,000 years, until someone discovered it and dug it up.

However, by that stage, the cycles of thawing and freezing had taken their toll. If you don't feel an immediate urge to put a mammoth steak on the barbecue, it's safe to assume that there's little chance of finding living or undamaged cells.

It is still possible to extract genetic material and piece it together after the cells have died and started to decompose. However, the problem is that the genes that scientists fit together exist only in a computer program. The analysis of hereditary material, the complex sequencing of minute fragments of DNA, is an entirely digital process. To be able to clone a cell, however, a complete, undamaged version of the DNA molecule is needed. So far, no scientist has found any mammoth cells so well-preserved that there is even the remotest possibility of cloning them. Yet there is another way.

Scientists are already able to construct tiny fragments of DNA and insert them into cells. This is the method that is beginning to look feasible in Boston, where I spend a few very rainy days.

George Church is a professor of genetics at Broad Institute, set up in cooperation between MIT and Harvard. He looks rather like a taller version of Father Christmas. Though he lacks the traditional paunch, he has a thick white beard and curious eyes. Just like Father Christmas, he gets letters from enthusiastic children, but, instead of wish lists, these contain questions about mammoths, which George is working to bring back to life.

Before I travelled to Boston to meet him, another scientist I interviewed said, 'People would call him an incurable, almost crazy, optimist, if it weren't for the fact that all the scientific progress he predicted has actually come true. And in many cases, it's actually happened in his own lab.'

He was one of the scientists who mapped the human genome. A few years later, he came up with a method for analysing DNA that is both faster and cheaper than previous ones. Now he is already some way towards building a mammoth.

'There are definitely no animals yet. There are some cells in a culture dish, and in practice they're still elephant cells with a few modifications. But we've made a lot of progress,' he says.

The dish George is talking about is in a fridge at the end of the lab. It contains a thin, translucent, reddish fluid. This liquid looks like much-diluted blood, but is actually the solution in which the cells are being cultivated. The round skin cells are clearly visible under a microscope. They come from an Asian elephant, but contain genes that are copies of a mammoth's genes.

'The mammoth and the Asian elephant are more closely related to each other than either of them is to the African elephant. Anything that's shared between the African and the Asian elephant is probably going to be shared with the mammoth. Cold-tolerance is the exception,' says George.

The whole thing began with the sequencing of the mammoth genome, in which George played a part. But there were also a lot of questions from journalists who were curious to know whether it might be technically possible to recreate mammoths. These questions set George thinking, and, after discussions with various researchers in the field, he decided to give it a try. He and his colleagues began to examine the mammoth's genetic make-up in an attempt to identify the genes that enabled it to live at 50 degrees below zero.

Once they had identified a number of genes that might be

those that gave mammoths their unique characteristics, the next stage was to construct synthetic copies of them. This meant taking information from a data file and translating it into an actual DNA sequence that can be understood and used by a cell.

The method they used is known as CRISPR/Cas9. Developed in 2012, it has revolutionised scientific possibilities in the field of genetic engineering in a number of ways. One of the major problems facing scientists when they attempt to splice new genes into hereditary material is the difficulty of ensuring that those genes end up in the right place. This means that they used to have to make multiple attempts before achieving the desired final result. CRISPR/Cas9 works like a pair of scissors that can cut genetic material more precisely and exactly, thus making it far easier to do things like inserting new genes in the right location. As a result, fewer attempts are needed, and experiments can proceed far more rapidly.

This method's potential is not confined to recreating extinct beasts. The general consensus, in fact, is that it will have most impact on medicine. George was among the first to demonstrate that the technique can be applied to human cells. It is hoped, for instance, that scientists may be able to cure certain diseases by editing a person's genes, or by extracting and modifying stem cells.

In the spring of 2015, Chinese researchers published a study in which they showed how this method can be used to genetically modify human embryos. They attempted to replace a gene that causes a serious hereditary blood disease. Although the experiment was not quite as successful as the scientists had hoped, the method clearly has huge potential. In the spring of 2016, scientists from the UK, Sweden's Karolinska Institute, and elsewhere were authorised to apply the method to study the human embryo and how it develops during its first few days of existence, provided that none of the genetically modified embryos studied are actually carried to term.

Many scientists, including George, have also debated whether there should be a ban on applying this technique to human beings before it has been studied in greater depth. In theory, it is just this type of method that could be used to develop what are sometimes called 'designer babies' — babies that have been genetically modified to have the exact characteristics their parents want. However, it is also a method that has already enabled scientists to research areas and issues that were once entirely beyond their reach. At the time of writing, there is an intense debate in the scientific world about how the technique can be used in an ethically defensible way.

At all events, regardless of its possible significance as a way of curing diseases in humans, this is the method that enables artificial mammoth genes to be implanted in cells from Asian elephants. Step by step, George is converting elephant cells into mammoth cells. So far, he and the other researchers at the lab have made a total of 45 genetic changes to the elephant cells, making them more similar to the cells of mammoths.

The purpose of some of the new genes is to give the animals fur. Mammoths had a dense, curly pelt comprising a layer of hair designed to trap warmth combined with stiff hair on the outside to keep out dirt and moisture. I saw a few tufts of mammoth hair in the Yakutsk Mammoth Museum. The hair of the mammoth concerned was like mine in colour — a light auburn — while other samples were almost black. The pelt was long, too, up to 90 centimetres at the sides. The tip of the mammoth's tail had extra-long hair, presumably so the animal could use its tail to swat flies in summer. It moulted in large tufts in springtime, just like most other animals living in a cold climate. To be able to reproduce these features, George has to identify all the right genes.

A few other genes that geneticists plan to modify have the function of providing elephants with subcutaneous fat, while others serve to make their ears smaller. The overall purpose is to improve the new mammoths' capacity to conserve heat. When George talks about the project, he sounds a little like Father Christmas handing out gifts to a future baby mammoth — features that will enable it to survive in the real world.

It's difficult to know exactly which genes code for which features. Scientists base their reasoning on what's known about similar genes in other species; they compare the genetic material they're working on with genes that determine the appearance of mouse or dog hair, for example. This is largely a matter of educated guesswork.

The modified cells also contain genes that alter the elephants' blood. Despite the mammoth's subcutaneous fat and thick pelt, the tip of its trunk and other parts of its body that were exposed to the cold could become so chilled that ordinary blood would not have been able to supply the oxygen it needed. So mammoths had a highly specific adaptation to their haemoglobin, the molecule in red blood cells that carries oxygen. George has managed to recreate this. This reconstructed haemoglobin is the only modification the scientists concerned have been able to actually put to the test so far. Another team of scientists has demonstrated that the synthetic genes modelled on the mammoth's genetic material do produce haemoglobin that functions at very low temperatures.

None of George's experiments with mammoth genes have yet been published in a scientific study. He says he prefers to wait for more results before taking that step. From a scientific point of view, this means it's impossible to say anything about either the experiments or the results for the time being. However, on the basis of George's previous breakthroughs, I choose to believe in him — even if I am mildly sceptical.

The fact that scientists have managed to create new variants of genes that have been dead for 10,000 years is impressive, but it is still a far cry from a whole mammoth that can run about and romp in the snow. The next step is to coax the modified cells into becoming stem cells. The cells in the culture dish are dividing and growing successfully, but it has not yet proved possible to get them to develop into skin and hair follicles, for example, so as to check what kind of fur is developing. Stem cells are cells that have not yet taken on a specialised role within the body; they are present in embryos and in adults' bone marrow, for instance. The way that scientists can convert ordinary cells into stem cells is — just like CRISPR — a new and revolutionary method that has rapidly become standard practice in biological research. Though it was only discovered in 2006, scientists have since managed to generate stem cells for nearly all the animals to which they have applied the technique, including humans. However, it is precisely elephants that have caused difficulties for George and his team.

They have not yet succeeded in converting ordinary cells into stem cells. This might be because elephants are such long-lived animals, with strong in-built defences against cancer. Given the similarities between cancer cells and stem cells, it could be the elephant's anti-cancer defences that are thwarting attempts to produce stem cells. George plans to continue experimenting with this technique.

Without stem cells, the project would be stymied. Once they exist, the plan is to get these cells to form different body parts. This is a field of research that is currently developing at incredible speed in countless labs worldwide. If it worked, it would be possible to grow hearts or kidneys for people needing a transplant — by using their own stem cells, for example. As far as the mammoths are concerned, it's an essential step in order to be able to check how well the new genes function, and to be able to start adding further genetic modifications.

'I don't know how many genes we'll have to modify before we get there. We hope it's less than all of them — altering 20,000 or 30,000 genes would be a big job,' George laughs when I ask him to estimate how many modifications will be needed.

It is not until scientists have succeeded in converting the cells in the Petri dish into stem cells and tested all the modifications they have accumulated that they can start to think about letting them develop further into embryos and ultimately into furry little baby mammoths. It is at this stage that the task of making a mammoth gets really tricky — but I will come back to that problem later. Even at this stage, the research is impressive enough to raise some intriguing questions.

If George succeeds, will the new creature be a mammoth or an elephant? Since it will bear only a few mammoth genes, it won't be a copy of any mammoth that ever lived; it won't be a clone. Essentially, it will be an Asian elephant — but anyone looking at it will see a huge, hairy beast, and presumably they will think of it as a mammoth.

When I ask George what the animal will actually be, he seems to want to have his cake and eat it. On the one hand, he says it will be an elephant fortified by the characteristics of a mammoth, and that the project may help protect modern elephants.

'It's actually about conserving the Asian elephant. If you think about it, you only need a handful of genes to make elephants cold-resistant, which would mean they could live across a much wider area.'

The International Union for Conservation of Nature (IUCN) describes wild Asian elephants as a threatened species. There are only half as many of them alive today as in the 1980s. One reason for this is poaching, but another is that the forests where they live risk being razed as farming expands. The solution to that problem, in George's view, is to make elephants more like mammoths so they can live in Siberia, which has fewer people and more room. Tweaking a few genes to give the species a new habitat wouldn't stop it being an elephant, in his view. He explains:

'There are people whose genes allow them to climb to the summit of Mount Everest without oxygen, unlike the rest of us. But we don't say that makes them non-human, we don't say they're no longer human because of those genes. The new cold-resistant elephants will be able to mate and breed with other elephants, so in that sense they'll be the same species. But the whole point of shifting them northwards is to avoid a situation in which they have to compete with farmers for land, as they do today. It's about giving them a new place — or an old place, really — to live.

'And anyway, elephants already like snow. In wildlife parks, they'll go out and roll gigantic snowballs that are bigger than humans, and they'll jump and break the ice on pools as children do. They can only tolerate the cold for about an hour at a time, but they have a lot of fun during that time,' George chuckles.

The idea of protecting elephants by adapting them for life in Siberia has, of course, been criticised. One reason is that it fails to address the underlying problem. The forests where Asian elephants live today are teeming with threatened species. Moving the elephants would just mean abandoning all the other species to their fate. My feeling when I hear George talking about how this would help modern elephants is that the idea seems rather out of proportion, like trying to kill a mosquito using a fully automated designer drone. It may be a great thing for the engineer who builds the drone, but that doesn't make it the most effective way of solving the problem.

And now we come to the other side of the coin. In his next breath, George says the animals he creates will be mammoths as well. If they are released in Siberia, the idea is that they should play the same role in the ecosystem as mammoths did 10,000 years ago. They will look like mammoths and live where mammoths used to live. They will fascinate people and inspire them to become involved, George hopes. This is where the handwritten letters come in. This is why scientists are offering to work on the project pro bono, just so they can be part of it. Nearly all of them are passionate about mammoths.

'This is about being inspired by mammoths and recreating something that resembles them. It's not about paying a debt or atoning for our supposed guilt in killing off mammoths.'

What George wants his project to achieve most of all is for us to start thinking about how we can use all the new genetic tools to protect species that are in the process of dying out. He is worried about the future and about how we can save the species we still have, but thinks the new genetic technology offers a possible solution.

'We're facing a situation in which we can not only halt the extinction of species, but actually reverse the trend. That makes it all much more inspiring. This is about creating new organisms, animals that are better adapted to today's environment.'

The small round cells are going on an exciting journey. Although he began with mammoth genes, George is considering the possibility of giving the new animals characteristics taken from penguins or polar bears, if these might produce useful features that the mammoth never developed through evolution.

'We might be able to do even better than the mammoth did.'

It is just that thought — that we humans could learn to make new wild creatures — which crops up again and again during the writing of this book. Talking to George makes me feel almost dizzy. His visions and his undisguised optimism make it quite hard to take him seriously. At the same time, he is clearly extremely well-informed and knows exactly what he's talking about. I am seduced to some extent by his optimism; who doesn't want to hear that everything will come right in the end? So I ask him whether he really believes the mammoth will come back to life.

'Well, since you haven't put a time limit on it, I think it's very likely. The amount of money it costs will get less and less, while our know-how gets better and better every year. I think it's very likely there'll be a cold-resistant Asian elephant soon.'

But when I ask him how long it might take for him to achieve his aims, his answer is vaguer. It's hard to say, and you always look foolish if you say something will take 100 years to develop, and it then takes only a decade. Given the incredible pace of technological development, all George can say is that it will take at least another five years.

The cells in the reddish liquid in the Petri dish are just the first step in this project. Many scientific breakthroughs will be needed before there will be a furry baby elephant for George to pat. Yet none of the scientists I talk to — not even the ones most critical of de-extinction — think it impossible for George to create something like a mammoth. Quite a lot, however, doubt whether there will ever be so many such beasts that they can be released and roam Siberia once more.

CHAPTER 3
Zombie Spring

Just suppose you had a time machine that allowed you to travel back through the ages, but only to times and places where there were no humans — where would you go?

Would you visit the deciduous forests of Europe, which abounded in giant deer and aurochs before the first Europeans set foot there? Would you take diving equipment with you and go back 550 million years, during the Cambrian explosion, to view the teeming life of the oceans, that fauna in which most of today's animals have their origins? Or put on a space suit with oxygen tanks and try to locate the point where life on Earth began, between three and four billion years ago? Would you stand on the shore and watch the first four-legged creature crawling over marshy land nearly 300 million years ago? Would you try to see what exactly happened when the forebears of human and chimpanzees went their separate ways about five million years ago? Or would you visit the prehistoric forests of eight million years ago to look at the dinosaurs?

If you had no time machine, but you could resurrect a single animal species from among the many that have died out in the course of Earth's long history, which one would you select?

For George Church is not alone. About ten projects involving attempts to revive extinct flora and fauna are currently under way. It's easy to write off the scientists concerned as unworldly dreamers, or as charlatans motivated only by cash and celebrity for something that can't be achieved in any case. When I started out on my journey, my inner cynic was quite loud and insistent, constantly sparring with the wide-eyed, fascinated ten-year-old in me. After meeting George, listening to his ideas and peering into the Petri dish full of cells, that cynic quietened down somewhat.

We already know of organisms for which death is not eternal. Scientists have managed to thaw out harmless viruses that are still living though they have lain sealed in ice for 30,000 years. There's even some concern that rising global temperatures may thaw out dangerous viruses as well and allow them to start spreading. Other researchers have succeeded in coaxing frozen plant cells taken from seeds that are also 30,000 years old into starting to divide and grow into plants bearing pretty little white flowers. Microscopic eight-legged tardigrades (also known as water bears) can reach a state of frozen suspended animation that enables them to survive desiccation, a vacuum, and extreme cold for a very long time.

None of these phenomena are comparable to the resurrection, de-extinction, or revival of extinct species that scientists are now seeking to achieve, using gene technology of varying degrees of sophistication.

'We [humans] are as gods and might as well get good at it.'

This is Stewart Brand's mantra, a slogan he coined back in the 1960s. I meet him in a cafe-cum-library in San Francisco, where he's running around in a baseball cap and a green quilted jacket. Stewart is well into his 70s and speaks with a distinctive drawl, adding a 'sh' sound to every 'w'. I hardly have time to get a cup of tea before he launches into some big ideas.

'There are three long-arc narratives I think are going to be dominating this century. One is climate change; another is urbanisation. The third, I think, is biology and biotech, which is developing at the moment the way digital technology developed 20 or 30 years ago.'

Stewart likes to think in terms of long periods of time, and he wants others to do likewise. He has already packed so much into his life that it is hard to introduce him in just a few sentences. In the 1960s, he was one of the movers and shakers in the environmental movement, while later he was involved in developing the internet in its early days. He has set up a slew of organisations, companies, and campaigns. In recent years, he has criticised the environmental movement to which he once belonged for its romanticising and doctrinaire tendencies. Conversely, he has been the target of criticism himself for having close links with environmentally destructive firms, as one of his activities is environmental consultancy.

In the mid-1990s, he set up the Long Now Foundation, which aims to get people to take a long-term view of humanity and the challenges facing it. He has reformulated his maxim about our godlike powers; now he says: 'We are as gods and have to get good at it.' In other words, we humans have to shoulder our responsibilities and find solutions to the climate crisis, environmental destruction, and accelerating species loss, for it is we who are the gods on this planet.

This also sums up his attitude towards species revival.

'That's emblematic of the current change in the approach to protecting species. We're going from defence to offence. We're taking action — trying stuff, experimenting, not just trying to conserve the little that's left using old methods,' he enthuses.

If anyone can be said to have launched the idea of de-extinction, it's Stewart and his wife, Ryan Phelan, the founders of an organisation called Revive & Restore. They held their first scientific conference on de-extinction in 2013. This was the starting shot that launched an organised movement and sparked off debate among the various scientists working in the field. Many of the experiments concerned have been under way for much longer, but it was Stewart and Ryan who coined the generic term 'de-extinction' and linked up the different projects.

Ryan and Stewart presented the revival of species as a way of making the world better and biologically richer. They speak about how resurrected animal species would herald a new chapter in the history of humankind. Just as when I talked to George Church about his mammoth cells, it is easy to get carried away by enthusiasm, by a sense of hope and the feeling that the world may become a better place. It's tempting to see the future painted in broad brushstrokes, in bright colours.

So let's take a few steps back. Is it really possible to resurrect extinct animals — before we even get onto the subject of entire species or ecosystems?

The answer to that question seems to be yes. However, it clearly depends on how the questioner defines 'resurrect'.

The mammoth is a good example. Unlikely though it is, someone might conceivably find some frozen mammoth cells in perfect condition and use them to clone a furry little mammoth calf. That calf would be a genetic copy of a particular individual that died perhaps 20,000 years ago. Since the ovum would come from an elephant, the new animal would not be identical to its ancient forebear, but most people would be happy to call it a 'mammoth' anyway.

So scientists would have a mammoth, which would be an amazing scientific achievement. However, that wouldn't be enough to recreate mammoths as a species, as the animal would lack a mate to breed with. The lonely giant would presumably live out its days in a zoo or a laboratory, perhaps with a few elephants as playmates. Or perhaps it would mate with an Asian elephant, producing calves that would be hybrids of two species, like a cross between a horse and a donkey. Maybe these hybrids would be able to reproduce in their turn, maybe not.

Attempting to resurrect a whole species rather than a single individual calls for a different approach. The mammoth that George Church plans to build using the cells in Boston won't be an exact replica of any mammoth that ever lived. He's trying to build a new mammoth, not to resurrect an ancient one. This implies that he could, in theory, create as many as he wants once he succeeds, as he can then repeat the process with cells from large numbers of other elephants. This would also mean that the mammoths would be able to breed together; if George first modified a cell from a cow elephant and then one from a bull elephant, the result would be two mammoths of different sexes. It would then be possible to breed a herd of mammoths — but in practice, they would be a new species, not replicas of an ancient one.

The same applies to the other de-extinction projects covered by Stewart and Ryan's umbrella organisation. All the scientists concerned are trying to create not just one individual, but a whole species. This is a sine qua non if the animals are to make the difference they hope to see. If they are to help create a better future, there must be significant numbers of them, and they will need to be released into the wild. But they will be new versions of their forebears, resembling the original species to varying degrees.

It will most likely be feasible to recreate animals modelled on an extinct species within a few years' time. The optimists think it will take less than a decade, the pessimists a good 20 to 30 years. It looks as if not even the critics discount the possibility. Exactly how it will happen, and what the scientists will need to do, varies from species to species. In a few cases, frozen cells from the extinct animals are available, so it may be feasible to attempt cloning in various ways. In other cases, it will be a matter of moving genes around and modifying a close relative. Species that disappeared a long time ago are not going to suddenly rise like zombies from their graves. We are not about to be startled by a stray mammoth while taking a morning constitutional.

'I think it'll come in stages; we'll add one trait after another from the extinct animals. There are so many gradations involved in all this,' says Stewart.

The genetic technology needed to achieve this is very quickly becoming both more sophisticated and more affordable. Stewart again draws a comparison with changes in digital technology.

'At the moment, it's being done by grad students with pipettes, but at some point it's going to be a robot. So if you ask that robot to change 14 genes, it'll say, "Why 14 — how about 14 hundred? Why not change the lot of them?" And you'll say, "Okay, change the lot!" "That'll cost you a bit extra," the robot'll say. "How much?" "Oh, that'll be 4,000 dollars." "Sure, that's not so bad, go ahead then!"' says Stewart, chuckling as he conjures up a conversation with a future robot.

The next question is what will happen once technology has caught up with the scientists' dreams. George is planning to release his mammoths in Siberia; likewise, all the other projects aim to release the species they are working on. All the researchers involved are working towards the goal of having their species live in the wild, independently of humans. (Dinosaurs — for obvious reasons — are an exception.) This means scientists are aiming not to create exact replicas, but to produce animals that are sufficiently close to the original to be able to represent the extinct species, creatures that can function more or less as their forebears did in their original environment. Moreover, they need to be able to live in the environments that exist today and to coexist with contemporary species.

More than any other factor, it is the aim of releasing animals into the wild that makes this field of research controversial. It is this that has attracted both praise and criticism. Most of the resurrected species will be genetically modified organisms, GMOs. What people would think about significant numbers of genetically modified wild animals being released is an open question — to put it mildly. Few people today are concerned about the fact that many drugs are produced using genetically modified organisms, but more are anxious about GMOs when it comes to carrots, potatoes, maize, and tomatoes.

'Ryan has worked in genetics for a long time, and she says you should never underestimate the amount of fear that people have of anything genetic,' says Stewart.

A few weeks after talking to him, I find a website with pictures of all the creatures that have died out over the last hundred years. The picture at the top shows the golden puma that once lived in the eastern United States, but was declared extinct in 2015. At the bottom is a painting of a brown and grey owl. When it was discovered, it was named the laughing owl because of its call, but it died out back in 1914.

The IUCN's list of extinct animals contains 866 species, from the aurochs, which died out in the 17th century, right up to today. It also has a list of a further 69 species that are extinct in the wild and which can now be found only in zoos. The disappearance of 900 species in the space of 500 years may not sound so very serious — particularly when you reflect that scientists have discovered about one-and-a-half million types of fauna. The exact numbers of species of fauna and flora are still unknown; the most recent major study suggests that they amount to just over eight million. Scientific estimates vary from a few million to as many as 50 million.

But the 900 animals on the IUCN's list are just the tip of the iceberg. I look at photos and paintings of rhinos, big cats, bats, tortoises, snails, and frogs, knowing that they represent no more than a tiny fraction of all the species that have actually disappeared. These are the species that meet two criteria: they were documented before they became extinct, and we can say with absolute certainty that there are no individuals still hiding in the depths of a forest somewhere.

Trying to estimate how many species have really died out since the 17th century is a tricky matter; again, scientists' estimates vary tremendously. Large numbers of species become extinct without humans even having noticed their existence in the first place. In addition to animals, many plants, fungi, algae, and other organisms have disappeared.

In most cases, the extinction of a species is a trivial matter. It happens all the time and is an essential part of evolution and the capacity of living things to adapt to new circumstances. In a changing world, not all species survive. Over 90 per cent of all species that have ever lived on Earth have died out, from the first floating multicellular organisms to the giant ground sloths of South America.

Moreover, a few times in the course of the history of living things, nearly all species have disappeared simultaneously. Some catastrophe has struck the Earth, fundamentally altering the conditions for life. The remains of such mass graves, visible among the fossils discovered by palaeontologists, show that there have been five such mass extinctions over the last 500 million years. The first of these took place 450 million years ago, when an estimated 70 per cent of all species disappeared. All the animals in the world lived in the oceans at that time, and it's unclear what exactly happened; one theory is that both the sea level and the temperature fell drastically for some reason.

The most recent of the five great extinctions took place about 65 million years ago. The Earth was struck by an asteroid, and 75 per cent of all species died out. It was then that the dinosaurs disappeared, apart from the group that evolved into birds.

Now, more and more scientists say we are living through a sixth period of mass extinction, a new global disaster and time of change. On this occasion, it is humans who are responsible for the meltdown. It is we who are making species disappear far more rapidly than they would have done without us. We have hunted them and changed their habitats, and sometimes the extinction of one species has led to that of another. According to one of the most recent scientific estimates, humanity is responsible for the loss of 13 per cent of the Earth's species over the last 500 years — more than a tenth of the total. Most have disappeared from the United States or Europe, the regions where we have had most impact on nature.

Scrolling down the IUCN web page with images of extinct animals, I reflect on how impossible it all seems to grasp, this mix of figures and attractive pictures that I can't quite get my head around.

There is an ongoing discussion about when this effect actually began. Many scientists claim that it goes back to the time when we began to kill mammoths and other prehistoric animals. Stewart is one of them.

'We humans have made a big hole in nature over the last 10,000 years. Now we have an opportunity to repair part of the damage. We need to find different ways to protect the species that still exist. But we can also recover some of the species that have disappeared.'

He has visions of an Earth that has gone further than just recovering a reasonable level of biodiversity, or what he calls 'bioabundance'. This isn't just about resurrecting species; it's also about giving the creatures alive today a better chance of thriving, not merely surviving.

'I want the cod in the ocean to be the size cod used to be, for example. People go to the national parks in Africa and look at savannah full of animals, masses of animals and different species. Europe used to be like that, North America used to be like that, even the Arctic had that wealth of fauna. That's my goal,' he says, trying to convey his vision of the 21st century as the green century, the age in which the frightening development we are witnessing today is reversed. He sees before him a time of restoration and reconstruction after the last two centuries of destruction. His commitment to the environment is clearly just as strong as it was in the 1960s.

Regardless of what one may think of Stewart's green visions, it's worth mentioning that he doesn't think they will form an obstacle to economic growth. Rather, he sees us humans gradually freeing ourselves from our dependence on nature, which will mean we can give it more space. The forests have recovered in the US and in Europe, he says; trees are taking over abandoned fields. More efficient farming means less arable land is needed in technologically advanced countries. He is convinced the trend will continue, even if the rate of change varies from one place to another.

This is an open question. Some studies suggest that Stewart is right, that globally we will need less land to grow food. That could mean allowing large areas to return more or less to a state of nature. All this, of course, is based on the assumption that agriculture as a whole becomes more efficient and that greater efficiency isn't cancelled out by other factors, such as the cultivation of biofuel crops, or feed for livestock reared for meat. Other researchers foresee trends moving in the opposite direction, and warn of the risk of running out of arable land worldwide.

The fact is, the forests have already begun to return in Europe, the United States, and parts of Asia. To take just two examples, France now has as much forested land as it did at the end of the 17th century, and India's forests have gradually expanded since the 1990s.

Abandoned fields and resurgent forests have given rise to another movement, analogous to de-extinction: reviving lost ecosystems. There are organisations that seek to 'rewild' Europe and North America, to restore the vanished wilderness in its entirety, rather than just particular species. These ideas have also found a place under Ryan and Stewart's umbrella. Stewart sees all these factors as being interlinked: the new biotechnology, the capacity to resurrect extinct animals, humanity's reduced dependency on nature, new wilderness, and the promise of bioabundance.

In the middle of our theoretical discussion, Stewart asks me how Swedish beavers are faring. Beavers died out in Sweden in the second half of the 19th century, but were reintroduced in the 1920s. Animals from Norway and other countries were released and have now established a new population.

'They're doing well,' I say, slightly taken aback by the question, and tell him there are now so many beavers that they are hunted, just like roe deer and elk. 'Great!' he says, beaming.

'I think a measure for success in projects like this will be when former game animals are once again game animals, when they become common enough.'

Stewart also hopes that the genetic tools being developed for de-extinction purposes could be used to save animal species on the verge of extinction today, for example by helping them avoid inbreeding and congenital diseases. It goes without saying that there will be some difficulty in getting this idea to work. It relies on developing new technologies, acquiring new knowledge, and maintaining people's commitment. Even under ideal circumstances, the methods involved are complex and labour-intensive, and the best they can achieve is something that resembles the lost species. And whatever happens, the technique can't be applied to most extinct species. However, he thinks it might be a necessary crutch for the species that are in the process of disappearing today, and a possible solution for a few of those we have already lost.

The world he envisages is both tempting and at the same time disturbing. It's a world where we humans have made ourselves the managers of nature, including that part of it that we today regard as wild and untouched. It's a world in which people assume responsibility, but also exercise power: releasing new animals, building new versions of ecosystems, editing animals' genes to enable them to thrive better. A high-tech, utopian *Star Trek* for biology. I am not quite sure what I think about the concept.

While I'm doing research and trying to digest our conversation, I come across the term 'solastalgia', a neologism coined by the Australian philosopher Glenn Albrecht. The term refers to the sorrow and melancholy people experience when their natural surroundings, in a landscape they love, undergo change: for instance, when forests, meadows, or lakes become unrecognisable as a result of human activity. I reflect on whether my feelings on hearing about Stewart's visions are a kind of premature solastalgia, reflecting all the changes that nature would undergo.

After listening to all these high-flown plans, I also feel that I need to see something concrete, to form a clearer view of what such a future might entail in practice.

CHAPTER 4

A Winged Storm

This book is a story about the last, and the first.

Martha is perched on a twig with her back to me, her head turned so I can see her clear red eyes and the feathers of her nape with their subtle rainbow shimmer. She is mostly brown otherwise, her plumage ranging through a variety of warm shades. Her tail is long and pointed, her body more slender than that of an ordinary urban pigeon. The male perched behind her is more eye-catching, with his peach-coloured breast and shimmering mauve throat, but it is Martha that holds one's attention.

Martha is dead. She has been dead for over a hundred years. On 1 September 1914, at one o'clock in the afternoon, the staff at Cincinnati Zoo found her lying on the ground in her enclosure. She had reached the age of 29. Her body was frozen in a block of ice and sent here, to the Smithsonian Institute in Washington, DC. Her intestines were preserved in formalin, and a taxidermist mounted the body in the graceful pose I am now contemplating.

'That was one of the few cases in which we know exactly when a species died out, right down to the hour,' says Christopher Milensky, who works with the Smithsonian's bird collection and is standing next to me looking at Martha.

Martha was the very last passenger pigeon. She had been

living in the zoo on her own for four years, although large sums
of money had been offered as an inducement to find her a mate.
It's not unusual for species to die out, and one individual had to
be the last of her kind. But Martha attracted particular attention
because only 50 years earlier the passenger pigeon had been the
world's most abundant species of bird.

It's impossible to know exactly how many passenger pigeons
there were, but their numbers probably stood at between three
and five billion in the eastern United States by the mid-19th
century. By comparison, today's Sweden is home to a total of
about 70 million birds of different species. The pigeons lived in
large, dense flocks, and there are descriptions of how they would
darken the heavens for three days at a time, their droppings falling
'like snow'. In their breeding grounds, their dung formed piles
up to 30 centimetres thick. They consumed everything in their
way, leaving trees stripped of fruit and nuts. They are even said to
have perched on each other's backs so they could reach the grain
when they landed in the fields. Each migrating flock numbered
hundreds of millions of pigeons, possibly even up to a billion. As
far as we know, there were fewer than ten huge flocks in the whole
of the United States.

These flocks were constantly on the move across country.
Unlike other migratory species, they had no set routes. They
would avoid places they had already visited for a number of years,
until the fruit and nut-bearing trees had recovered from the last
invasion. That meant it was impossible to predict when they
would turn up, and only such people as happened to be in the
right place at the right time could shoot them. Passenger pigeons
were hunted intensively, first by the Native American population
and later by the European settlers. The birds' strength lay in the

density of their flocks; the numbers killed were never such as to pose a threat to the species.

Passenger pigeon (*Columba migratoria*), by John J. Audubon, Pennsylvania, 1824.

As Christopher explains, it was the telegraph and the railways that did for them. The telegraph enabled the flocks to be located, while the railways made it possible to speedily dispatch dead pigeons to the cities, where they would be eaten. Once people knew the flocks' locations, hunting them was like shooting fish in a barrel. They were trapped using large nets, and there are descriptions of how people killed them by standing on a hill and waving a stick about in the midst of a low-flying flock. The bodies were transported in barrels, on ice; for a while, passenger pigeons were the cheapest source of meat in the United States.

Now the flocks were dwindling fast and voices were being raised, saying that passenger pigeons were disappearing and needed protection. But this concern was ignored, not just because pigeon meat was so important to the economy, but because the very notion that human beings might exterminate a species seemed crazy.

Hunters shooting at a flock of passenger pigeons in Louisiana. From *The Illustrated Sporting and Dramatic News*, 3 July 1875.

At the turn of the 19th century, the list of animals that humans had caused to die out was still reassuringly short. The best-known of this select group was the dodo, another member of the pigeon family, which lived on Mauritius and was discovered by sailors in the 16th century. Larger than turkeys, fearless and flightless, they provided a welcome additional source of meat for vessels passing the island. Later on, live dodos were taken to menageries in London and elsewhere. Their numbers fell swiftly. The last time a dodo was sighted was probably in 1662, but it would take until the 19th century for even scientists to fully grasp that it was extinct. One reason for this was that many people, including scientists, soon began to believe that the dodo was a mythical creature that had never existed in real life. It was popularised in art and in literary works, such as *Alice in Wonderland* with its pontificating dodo.

But above all, there was a conviction that it was impossible for creatures to die out. This idea was based partly on religious arguments. It was, quite simply, inconceivable that a God-given species should disappear; moreover, nature was believed to be extremely stable and static. The first person to prove that animals really had died out was the French zoologist Georges Cuvier, who excavated fossils and showed how they fitted into the Linnean classification system and how they were related to species still in existence. In a groundbreaking article written in 1796, he proved that bones from mastodons (relatives of the mammoth) must have belonged to a species distinct from elephants. This meant there must once have been species that were now extinct. Hitherto, fossils had been explained either as creatures that had lived in antediluvian times or as variants of species now living in other locations. Yet species extinction was still viewed as an exceptional circumstance. At the end of the 19th century, even scientists believed there were still mammoths living in the remote wilderness of Alaska. So maybe it was not so strange after all that no one took the warnings about the dwindling population of passenger pigeons seriously.

The last wild passenger pigeon was shot on 22 March 1900 by a boy with a BB gun. The only passenger pigeons left were the few individuals living in zoos worldwide, of which Martha was the last.

'The passenger pigeon was a "superbird" that created its own ecosystem,' says Ben Novak. 'They were a winged storm that swept through the landscape, with the same effect as forest fires. They flew around in their gigantic flocks for thousands of years, and they'd still be doing that if it wasn't for us.'

Ben is determined to resurrect the passenger pigeon, and he plans to devote the rest of his life to achieving that goal.

'People need to start looking at the possibility of species resurrection in the same way they see the space race. This is going to take a long time, and it's about driving technology and science forward in giant strides.'

Born in 1987, Ben Novak is young and looks like an archetypal Californian hipster. He sports a neat goatee and his desk is scattered with plastic Transformer figurines. I meet him at his laboratory in Santa Cruz, a city not far south of San Francisco whose population seems to consist entirely of hipsters, hippies, and surfers. All the food in the local grocer's is organic and produced by small local firms. There are signs on the beach proclaiming, 'There is No Poop Fairy', to remind dog owners to clean up after their little darlings.

A little tired of the hippy vibe, Ben jokes that the latest universal trend here seems to be imagined gluten-intolerance. The only obvious thing that differentiates him from most other young researchers at the university is the fact that his great passion is pigeons.

'My granddad taught me to raise pigeons and feed them by

hand. My dream is to have a big jungle-like enclosure full of exotic species, where I can have my morning cup of tea or a glass of wine in the evenings.'

At 13, he decided he wanted to bring back the dodo, but a year or so later he saw pictures of the extinct passenger pigeon in a book and decided that was the species he was going to revive.

'I saw a stuffed passenger pigeon for the first time when I was 16, and that was really, really cool,' he says. 'To begin with, I thought of de-extinction as a way of recreating the past, but the more I think about it, the more it's about humans accepting responsibility and helping nature to recover from the depletion it's suffered so far.'

The main reason why Ben wants to have passenger pigeons back is the very characteristic that many people see as the species' greatest drawback — the fact that they swept across the country like storms, sometimes destroying everything in their path.

'A forest needs a forest fire now and then for all its species to survive; that's how they've evolved. Similarly, forests in the eastern United States adapted to being struck by these flying storms from time to time. For example, oaks would do better and they'd produce more acorns if there were passenger pigeons shaking their branches now and then.'

How will Ben bring these birds back to life? The first step is to examine genetic material from passenger pigeons, with the help of hundreds of stuffed specimens in museums worldwide. Incidentally, he aims to see every single one in existence, and was enthusiastic when I told him there were two on show at the zoology museum in Lund, Sweden.

He has taken samples from the feet of a number of the world's stuffed pigeons. The samples come from the fleshy pad on the inside of their claws, more or less where you might pinch your own finger; the bit I imagine the witch in the tale of Hansel and Gretel pinched to judge whether the children were plump enough.

It's hard to examine the genetic material of stuffed specimens that have stood gathering dust in a museum for the last hundred years. Apart from the genetic intermingling and degradation common to all dead animals, such specimens are treated with various chemicals to preserve them in better condition, and these often cause the DNA molecule to disintegrate even further. The toe pad on the inside of the claws is the part where genetic material survives best, although it is quite fragmented even there.

Ben has used the same method applied by other scientists when piecing together mammoth DNA — taking DNA from a closely related species as a model. This is akin to doing a jigsaw puzzle by comparing each piece of the puzzle with the picture on the lid. Having compared the genes of the passenger pigeon with those of its closest living relative, the band-tailed pigeon, Ben has succeeded in piecing together all of the passenger pigeon's DNA.

'The next step is the fun bit — working out the differences between the species.'

The idea is to transform the band-tailed pigeon little by little into a passenger pigeon, just as George Church in Boston is converting an elephant into a mammoth. To do this, researchers will first need to identify the genes that made the passenger pigeon unique. Some of these relate to its appearance. The passenger pigeon had a long, wedge-shaped tail, which played a vital role in enabling it to fly at speed. Another aspect is that male and female passenger pigeons looked different, whereas male and female

band-tailed pigeons are almost identical. Yet even if Ben and the
other scientists working on the project were to produce a pigeon
that looked exactly right, that wouldn't be enough in itself.

'It's their behaviour, their propensity to live in dense flocks,
that's key in this experiment. If we can't get these pigeons to form
flocks, we'll have failed,' says Ben.

The really tricky part is thus to identify the various genes
that determine the pigeons' behaviour. This area is far less well-
researched than the question of which genes determine an animal's
physical appearance. Numerous research teams worldwide are
trying to discern links between particular genes and the behaviour
of various species of animal, but the field is still in its infancy.
The next problem is that a gene that affects a particular type of
behaviour in mice may — or may not — resemble the gene that
influences the same behaviour in birds. However, Ben hopes they
will nonetheless be able to identify such similarities.

'We don't know exactly which genes determine the shape of
the pigeon's tail, the colour of its feathers or its neurology and
behaviour. This would have been an incredibly difficult task even
if we were talking about humans, and we know a lot more about
them. This is much trickier.'

He's trying to identify likely candidates — genes that might
have the right effect — and then to examine what happens if
they're edited. The next step will be to identify the combinations
of genes that interact with each other, which can then be edited
in the band-tailed pigeon.

'In a way, we'll be making a new species by taking genes from
an extinct species and inserting them into a living one.'

The really difficult part — actually editing the bird's genetic
material — will come once the team has identified the genes they

need to alter. Although it's now relatively easy to edit the genes of mice embryos in their early stages, for example, or genes in cells taken from an elephant, the technique is far more difficult to apply to a chick embryo inside an egg. Have you ever wondered how it is that the yolk of an egg lies precisely in the middle of the albumen, at exactly the right distance from the shell? Neither had I before I met Ben.

Unlike a foetus, which lies still in a stable uterus until it comes to term, he explains, a chick embryo and the material that will form the egg go on a kind of rollercoaster ride inside the bird. After mating, an embryo forms and lies like a speck of dust within the yolk at the further end of the mother bird's long oviduct. It then begins a helter-skelter journey through the winding oviduct, during which the egg acquires layer upon layer of albumen, before being spray-painted with the calcium that forms the shell and, finally, being laid by the bird.

By the time the egg is laid, the chick embryo is already so well developed that it's too late to edit its genes in any way that would affect the chick as a whole, as it already comprises thousands of cells. And there's no point in the egg's meandering journey at which scientists could operate on the mother bird, extract the egg that is in the process of forming, edit its genes, and put it back inside its mother. That would damage the egg and, presumably, kill the mother.

What they can do, however, is modify the eggs inside the chick embryo, which it, in its turn, will lay one day. While the embryo is developing, the cells that will eventually become eggs or sperm are housed in a particular area on one side of the embryonic chick. They will later move further inside the embryo's body and settle in embryonic testicles and ovaries. This means they lie near the surface and are readily accessible for a short period after the egg has been laid.

The scientists plan to extract some of these cells and implant genes from passenger pigeons, before re-implanting them and allowing them to develop. The modified pigeon chick, or squab, will look like any other young band-tailed pigeon, but will carry genetically engineered eggs or sperm within its body. If two such birds mate, their offspring will be a passenger pigeon — at least, according to all the criteria that count for Ben.

'It's their characteristic features we're after, even if we're going down the genetic route to achieve that. We don't want to create an exact copy of a specific passenger pigeon; what we want is to produce a bird that can fulfil the same role in nature as passenger pigeons.'

Various scientists have already succeeded in removing these cells from birds and re-implanting them, even in different species. They've managed to get hens to lay eggs that have hatched into ducks, quails, or guinea-fowl. But no one has ever conducted such experiments on pigeons, and the very first study to attempt genetic editing in this sort of cells took place in 2016. Developing the technique is going to be the biggest challenge facing the project.

'If it doesn't work, we won't be able to make any progress at all. It wouldn't matter how much we studied genetic material or how much we thought about which sequences we'd like to edit,' says Ben.

If they are successful, several generations of pigeons bearing one or more genetic modifications will be hatched, thus enabling the scientists to test the impact of these modifications in their search for the right combination of genes to create a bird resembling a passenger pigeon.

'We haven't picked a name for the first one yet. The lab boss wants it to be called Ed, after her boyfriend, but I'd prefer something more impressive. Anyway, we won't be naming the first one either Martha or George, like the last male,' says Ben.

It will take a few more years yet. Ben hopes that the first chicks with a single genetic modification will be hatched soon, and that there will be genetically complete passenger pigeons by 2022, ten years after the start of the project. But even if a squab is hatched with the right genes, that doesn't mean the project will have reached its conclusion. It is at least as important for the squabs to be raised in the right way.

'Genes represent the limits of what an individual can become, the outer framework. But it's the environment that determines where you actually end up within that framework. Just giving them the right genes to steer their behaviour doesn't necessarily mean they'll behave in the right way. That's where you get into that interesting interaction between environment and genetic make-up.'

So even after the first new passenger pigeon has hatched, several decades' work will be needed to find the right way of raising the squabs, to gradually get them used to forests and, ultimately, to release them into the wild.

'I don't have any other major life projects that would clash with this one. I'm planning on devoting the rest of my career to passenger pigeons,' says Ben. He laughs when I ask whether he feels at all anxious at the idea of having his whole career plotted out in advance. 'That's how I am, and that's what I want to do. It's actually pretty cool to be sure of having a job in future,' he says, smiling. He has a fiancée, and he jokes that the only reason she wants to be with him is the millions he's going to make from pigeons. That, of course, is irony. Ben is fairly sure he'll never earn a dollar from pigeons, and he hopes the project will continue to be managed by the university.

But it is here, at the point where discussion turns to actually releasing the pigeons, that much of the criticism of the project and many of the concerns that make people uncomfortable about de-extinction come in. What Ben is proposing, in effect, is releasing genetically engineered animals into the wild for the express purpose of having a major impact on the natural environment.

'The debate about GMOs, genetically modified organisms, is still quite new in many respects, and it's pretty heated and wound-up. We're going to manipulate and engineer the animals we create — that's what makes this different from other species conservation projects.'

As he sees it, the key difference between his attempt to revive the passenger pigeon and many of the genetically modified crops used in farming is that his project is not for profit. No one will be patenting the passenger pigeon's genetic modifications or making any money from releasing the pigeons into the wild.

The use of GM crops in agriculture has a chequered history. The first genetically modified plant was created in 1983, by which time scientists had already succeeded in editing the genes of both bacteria and animals. The method is used in a wide variety of research projects, and many medicines are produced by genetically modified organisms. Most of the insulin used by diabetics is produced in this way, for instance. But it is agriculture where the technique has had the biggest financial impact. Towards the end of the 1980s, field experiments were carried out with tobacco plants that produced a substance toxic to insects with the help of a gene derived from bacteria. This toxin protected plants against attacks by insect pests. It is called Bt and is still a common genetic modification in crops. Today, for example, much of the cotton grown in India bears this gene. One problem, however, is that insects have developed resistance to Bt, so the modification is no longer as effective as it once was.

In the early 1990s, a small number of genetically modified plants were approved for commercial cultivation: a potato, a tobacco plant, a tomato, a variety of maize, and others. Their use exploded, and today genetically modified crops are grown on 12 per cent of the world's arable land. Many of today's GM crops are resistant to the strong herbicides used to kill weeds.

There has been massive criticism and fear of this development. There is a broad consensus within the scientific community that eating GM food is not dangerous in itself, but also that the technique has been widely used to enable farmers to use larger quantities of more hazardous herbicides and insecticides. Companies such as Monsanto sell both GM seed and herbicides or pesticides that can be used in tandem with the seed. Moreover, there are studies suggesting that certain crops have adverse effects on insects, for example. The main risk, in many people's view, is that cultivated plants could cross with related wild varieties, thereby enabling the modified genes to spread into the natural environment and escape human control.

Another toxic issue around GM crops has been patenting, with firms that develop new GM crops patenting both the plant and the modification. This makes the crops costlier for farmers, and Monsanto and other firms that hold many such patents have been criticised for exploiting poor farmers in developing countries. In recent years, however, many of the existing patents have expired, so generic GM crops are beginning to appear, no longer owned by the companies that first developed them.

The possibility of genetically modifying plants also means that there's considerable potential for developing crops that could result in more environment-friendly farming, using fewer resources. There are experiments under way that involve splicing genes into plants to make them more resistant to drought, cold, salty soil, and other phenomena, though far less progress has been made in this area. The most widely grown crop is golden rice, a variety developed in 2000 as the first GM crop with an enhanced nutritional content. Golden rice contains extra vitamin A to combat the deficiency diseases affecting many children in poor parts of the world where rice is the staple diet. An estimated 600,000 children die every year from vitamin A deficiency. However, even golden rice has been criticised by organisations that oppose GM crops.

Ben and I sit discussing the risks of genetic modification and people's fears. He hopes his pigeons and other revived animal species will avoid these problems, precisely because there is no commercial interest at stake. There are no big corporations involved that are keen to make money, and the aim is to care for the natural environment.

'I think it's very important, both for myself and for society, that these first de-extinction projects are focused on restoring ecosystems; it's important that they're not about producing pets or animals for lab experiments. I wonder whether people would have a different view of GMOs today if the first products launched had been vitamin-enriched rice, not plants that are resistant to extra-strong pesticides.'

Another question he faces when he discusses the project

with members of the public is a more philosophical one. Would resurrected animals be monsters, since — unlike other animals — they would not be the outcome of an evolutionary process?

'At that point, I always feel like asking people to take out their mobiles and look at the photos they have of their pets — especially cats and dogs. These pets are animals that have no counterpart in nature, animals created by us, manipulated and selectively bred to bring out the characteristics we love in them.'

Once the scientists have succeeded in producing passenger pigeons, the first few can be moved into a large enclosure in the woods, so they can get used to the natural environment. The next step will be to get them to form flocks that fly between various locations. In this respect, passenger pigeons' behaviour makes the scientists' lives a little easier; the parents used to leave their young before they could fly.

Imagine a whole tree full of birds' nests and squabs left to their own devices, not yet quite able to fly. They call to each other, beginning to form friendships. Before they've even taken flight for the first time they form what scientists call a proto-flock, a group that will stay together for the first part of their lives. They learn to fly together.

'What we think they used to do — and this may be quite decisive in terms of their social development — is form their own flocks as youngsters, after which they'd join a flock of adult birds once they were strong enough. So the social ties between young birds in a group could be very important indeed,' says Ben.

After a while, a group of adult passenger pigeons would come flying past, and the little flock of youngsters would be drawn by their flapping wings and take flight. To encourage this behaviour, Ben wants to train groups of carrier pigeons to fly from one spot to another, dye their feathers to make them look like passenger pigeons, and have them fly past the young pigeons. That would enable the youngsters to learn to fly about, and using carrier pigeons would enable the routes to be varied, to prevent the young birds from beginning to follow set patterns. Then they could gradually remove the carrier pigeons and let young pigeons fly with older birds that had already flown a number of routes.

'When we think the birds are flying in the right way, that they're flocking together closely enough and moving in the right way, the idea is that we'd remove the nets and let them start to discover the world and do their own thing,' says Ben. That's the plan, though no one knows how many pigeons there would have to be in a flock for the young birds to want to follow them, or for adult birds to feel secure enough to start laying eggs and reproduce. Passenger pigeons relied on having large flocks; it's possible that thousands of birds will be needed before a re-engineered pigeon can feel at ease.

The final goal is to have such a large flock that the pigeons will start to have an impact on the forest again, and to change it. Ben thinks at least 100,000 birds will be needed to have an ecological impact, maybe even as many as tens of millions. There's no way of knowing for sure.

'I think the forests could cope with a flock of a billion birds again. There's plenty of food in the forests, and plenty of room for a new passenger pigeon. There are lots of trees over a hundred years old, trees that remember these birds and have felt their feet on them,' he says.

How would that be, I wonder. If great flocks of pigeons began flying around the United States again, if there were so many that they started to have an impact on the forests, to surge onward like a forest fire of empty nutshells, broken branches, and piles of bird dung. Yet if there were so many that they genuinely played a role in nature, there could also be enough to pose problems.

'What we want them to do is something that might scare people — because we want them to disrupt and destroy parts of the forests, just like a heavy hailstorm or a forest fire. But it would be a disruption that wouldn't involve the same huge risks as fires, for example. Sure, it might be a real nuisance to get your car or your house covered in bird droppings, but it would be very different from having a fire in a residential area,' says Ben.

I have a feeling he's being just a little naive about what people will be prepared to put up with for the sake of these pigeons. Preventing people from exterminating them again may actually prove to be a greater challenge than all the work on genetics. In the course of our discussion, he paints a picture of a possible future.

'With other kinds of pigeons, you only see one or two at a time, though there may be 300 million of them in the whole country. There'd be that number of passenger pigeons in a single flock, which would be confined to a few square hectares at any one time. If you were in the same place as the flock, you'd have a storm of birds, but everywhere else it'd be as if they didn't exist.

'Just imagine a flock of ten million birds landing in the woods outside New York in 2085. Everyone would be fascinated by such an incredible number of pigeons and the effect they'd have on nature. Birdwatchers and school classes would come to see them. There wouldn't be any passenger pigeons anywhere else in the United States for those months; they'd all be concentrated within a small area. Then they'd move on somewhere else. They wouldn't be back next year, or the next, or the next after that. It might be five or ten years before they returned to the same place. So the disturbance would be intense, but short-lived.

'But we don't know how many birds we'd need for there to be this positive impact on the ecosystem. What we need is to find a balance between our human needs and ecological needs — and a robust, healthy ecosystem gives an incredible amount back to the people who live in it.'

I'm tempted; it sounds as if it would be truly awe-inspiring to see such a huge flock of birds. Maybe it would be as breathtaking an experience as seeing Mount Everest, or gorillas in the wild, or a pod of killer whales off the coast of Norway. Maybe it would be so magical that no one would mind having their car covered in bird dung, or their favourite trees in their gardens destroyed.

'Our aim is to get people to think less about how it would affect them, and more about how it could affect the world as a whole. Seeing the positive effects of a species brought back from extinction could be even more significant for what we do in society and how we look after our environment than the tragedy of a species dying out. It could have a significance beyond just this one species.'

CHAPTER 5

New Kid on the Block

It is 30 July 2003, and a caesarean section is being performed in an operating theatre. A handful of people in blue scrubs, wearing white plastic gloves, surround the mother. Alberto Fernández-Arias pulls out the newborn, a fine-looking kid with short, thick, grey-brown hair. She measures nearly 50 centimetres from her dark muzzle to the tip of her short tail. Her legs are long and slender, her hoofs white.

This is the first ever instance of species resurrection. The little kid is a clone of a female who died three years previously: the last *bucardo*, as the Pyrenean ibex is known in Spanish.

The mountains of Spain have always been home to ibex (mountain goats), the largest being the bucardo or bucardón, which lived in the Pyrenees, near the French border. Owing to their long, curved horns, they became hunters' trophies early on, and they are depicted in 15th-century paintings. These ibex could literally climb walls. They wandered about on near-vertical rock faces, leaping deftly over precipitous drops. That characteristic made them both harder to hunt and more prized as quarry. They were already starting to become rare by the 18th century. This, of course, simply boosted their hunters' prestige still further.

By the beginning of the 20th century, the bucardo seemed to

have died out completely. No hunters had spotted any for several years. At the same time, the scientific world began to take more of an interest in the Iberian ibex, subdividing them into four subspecies, of which the bucardo was one. Another subspecies, the Portuguese ibex, had already died out by the end of the 17th century. But the Pyrenees cover a large area, and a small group of bucardo was finally discovered in a remote part of Ordesa in north-eastern Spain. The mountains where they lived were soon made a protected area. In 1913, it became illegal to hunt the bucardo, and this small group of animals survived, though it never expanded beyond about 40 individuals.

Alberto Fernández-Arias comes into the picture in 1989. In that year, he qualified in veterinary science, with a particular focus on wild animals and reproduction. In the course of his training, he had treated many injured animals, ranging from bears to eagles. During his compulsory national service, he was tasked with developing methods for the assisted and artificial insemination of the bucardo, which the Spanish authorities wanted to try and save. Research indicated that there were only between six and 14 individuals still alive.

'When I started, we knew nothing, not even the simplest things which you would have thought would already have been studied by someone. We were the first to do so much of what we did.'

The first problem was trying to catch the creatures. Alberto didn't want to risk injuring the few specimens still living in the wild. Instead, he decided to conduct all his research on a relative of the bucardo, an ibex from southern Spain. As they were starting from scratch, and ibex are so good at climbing and leaping, it took several years to develop traps that were both effective and safe.

Illustration from *Wild Oxen, Sheep, and Goats of all Lands, Living and Extinct*, by Richard Lydekker. Illustration by Joseph Wolf, 1898.

The last time anyone spotted a male bucardo was in 1991, two years after Alberto's team began their research. As more time passed, all hope of capturing the remaining wild individuals and starting a breeding program seemed lost. But Alberto thought it might be possible to cross females in captivity with a closely related species and thereby save the bucardo. No one knew exactly how many females were left in the mountains.

'We were working under huge pressure to develop these methods, and, at the same time, every step along the way took such an incredible amount of effort.'

Once they had learned the right way to catch the ibex, and how to keep them in captivity without their injuring themselves or bolting, the scientists started hormone treatments to increase ovulation. They continued to work with the bucardo's close relative from southern Spain so as to avoid risking the few bucardo still living wild in the mountains. The scientists dared not catch them yet. The idea was that mature ova would be removed from these other ibex, fertilised, and then implanted into surrogate mothers — ordinary nanny goats — which would give birth to kids. As a result, each female ibex would produce far more offspring than she could otherwise have done. She would continue to ovulate, instead of falling pregnant and having to suckle and care for a flock of frolicking kids.

This sounds like a complicated method, but it is currently being tried out on many other species. For instance, scientists hope to be able to support the mouflon, the forerunner of modern sheep, in the same way. However, the technique was unsuccessful with ibex, resulting in numerous miscarriages and in offspring that apparently suffered from arrested development. Listening to Alberto describing the years of research and failed experiments, with one baby ibex after another being stillborn, I begin to reflect on the issue of suffering. How much suffering can reasonably be inflicted on individual animals in the interests of saving or reviving a species?

There's no doubt that Alberto is firmly on the side of animals and nature. Today, he's in charge of conservation in a region of

Spain. He has devoted his life both to the bucardo and to trying to protect much of Spain's natural environment. That makes the issue of suffering so much more complicated. In the course of our conversation, for the first time, I form a clear picture of the price to be paid for resurrecting a species. It's an issue that will crop up time and again.

While Alberto was developing the surrogacy technique, the animals still living in the wild disappeared one after the other. By the time the researchers had finally succeeded in getting the goat surrogates to give birth to healthy ibex kids, there was only one bucardo still alive. It was the late 1990s by then, and a decisive event had taken place.

'Dolly the sheep had been cloned. Up to then, cloning hadn't even occurred to us. Everyone had told us it was impossible, but now we began to think about it as an option.'

The world's most famous sheep was born on 5 July 1996 in Scotland. Dolly was not the first animal ever cloned — far from it. What made her truly unique was the fact that she was a clone of an adult animal; all previous experiments had been conducted on embryos. Scientists had taken a cell from a full-grown sheep, removed the cell nucleus containing the genetic material, and inserted it into an ovum. The result was a cuddly, photogenic ewe called Dolly, who was presented to the world in February 1997. Up to that point, most scientists had believed such a thing to be quite out of the question. Dolly graced the front page of many newspapers, bringing cloning into the public eye, and she played at least as significant a role in the world of science.

For Alberto, she was the hope that inspired him to continue his work.

In 1999, the Spanish authorities decided that the last bucardo was to be caught so scientists could take samples and preserve cells from her. During the same period, Alberto was also attempting to catch some of Spain's last bears, to implant radio transmitters.

'At night I was busy catching bears, while during the day I was trying to trap a bucardo, with a two-hour drive between the two national parks. It was terribly stressful.'

The trap the scientists had set for the bucardo in the mountains was the size of a skip. They lay on their stomachs some distance away, with their binoculars trained on it. On 20 April 1999, when there was still a little snow on the slopes, the trap fell shut. The last female bucardo was inside. The scientists went up to the trap, and Alberto took out his homemade blowpipe and some darts tipped with anaesthetic. He blew a dart at the ibex and waited for the anaesthetic to take effect before they went inside.

'We took two skin samples, one from the tip of her left ear and one from her left flank. We put a collar with a radio transmitter on her and took blood samples. Then we just sat there, in silence, inside the trap, waiting for her to wake up again.'

They named her Celia, and as soon as she awoke they released her again. She lived for a further ten months.

'She carried on living in the same area as before. But in January 2000, we could hear that the beeping from the collar with the transmitter had changed. When an animal's okay, it beeps intermittently, but, if something's happened, it starts beeping much faster. So we started searching. She was found crushed under a fallen tree.'

I can tell he finds it hard to talk about this. His voice is halting, and he has to swallow a couple of times. But when I ask about his feelings, he brushes my enquiry aside and says she was so old anyway that it was bound to happen sooner or later. Instead, he asks if I want to know why they called her Celia.

'It happened this way. The day after we'd caught her, I went to a meeting with the authorities to tell them how it had all gone, and I was interviewed by some journalists. My girlfriend — now my wife — and I hadn't seen each other for several days because I'd been up in the mountains, but she was with me that day. So when the journalists asked me what the bucardo was called, I looked at her and said her name. And the next day, it was in the papers — "Success for Operation Celia!" A bit later, my girlfriend's mum rang her and said, 'It's a funny thing, but that ibex has the same name as you." Her mum hadn't yet met me at that point — she didn't even know I existed.'

After 5 January 2000, the only living part of the bucardo that remained was the cells the scientists had frozen, so Alberto started trying to convert them into a living animal. The first stage was transferring the nuclei from the frozen cells into ova from ordinary goats, then allowing them to develop into the early stages of embryos. The researchers made over a thousand such embryos, and 154 were implanted into 44 goats, which served as surrogate mothers.

This is one of the major problems with all attempts to clone animals: it takes many, many attempts, and few of the cloned ova develop into embryos. Dolly, for instance, was the sole success out of 277 embryos. The next problem is when scientists try to implant the tiny embryos into their future surrogate mothers. To enable the embryos to develop normally in the surrogates' uteruses, Alberto had to develop a new breed, a cross between an ordinary goat and a relative of the bucardo — and even then there were large numbers of unsuccessful attempts. Cloning cells from old animals — Celia was probably aged over ten — is even trickier.

This is also the biggest difficulty facing George Church in his efforts to build a mammoth. The problem lies with the elephants that would be the surrogate mothers. It's quite a complicated process for elephants to fall pregnant, and they're prone to miscarriages if anything goes wrong. She-elephants, or cows, become pregnant only once every four or five years, and gestation takes over 600 days. There are quite simply not enough Asian elephants available for it to be feasible to make the hundreds of attempts that would be needed to bring a healthy mammoth calf into the world. There would also be major risks of injury and suffering for the elephants involved if the pregnancy didn't go according to plan.

So George has decided to try to avoid the problem altogether, by developing an artificial uterus.

'One of the biggest breakthroughs we would need, I think, is to get embryos to develop to term in the lab. That would take a huge amount of pressure off the threatened species. We could create an alternative way of producing new calves without having to bother the existing elephant population.'

He thinks this would be feasible; the plan is to develop an artificial umbilical cord to supply blood and nutrients to a foetus lying in a tank full of a liquid similar to amniotic fluid.

'It hasn't been done with a mammal yet. The way the science is today, you can get embryos to reach a given level of development outside the body, but you can't go to term, not even with mice. So we'd probably do it with mice first, then with elephants, and it's hard to predict how long it would take to develop the method fully.'

I am very sceptical about these possibilities, and about how easy he makes the whole process sound, despite the difficulties. Foetal development is a complex process about which much is still unknown. Surely large amounts of hormones and other substances would have to be supplied to the foetus at the right time — and scientists may not yet know when that is.

'Actually, we're not sure that's how it works. If it were a clock we were going to build, you'd need a lot of intricate handiwork to get it going, but biology is more about setting things off in the right direction and letting them get on with it. For instance, if you just put eggs and sperm together in a test tube, they'll form an embryo — you don't have to micromanage the process, just provide the right general conditions. There are lots of unknown factors — and that's true of all mammals, not just elephants. But that's fine, it means we'll learn some biology along the way,' says George.

George Church is clearly a born optimist, but I remain unconvinced. I think he's fundamentally underestimating the degree of complexity. However, it will be fascinating to see what he comes up with over the next few years.

So far, the artificial wombs are a long way off. Of the 44 nanny goats that had Celia's clones implanted in them, only one made it all the way through pregnancy without a miscarriage. It was time for a C-section in the operating theatre. In the Spanish lab over ten years ago, there was an intense silence when Alberto removed the little kid from its mother's belly.

'I picked her up and saw straight away that something was terribly wrong. She wasn't breathing as she should have been. We did everything we could to save her, but she died anyway. It took about ten minutes. I don't know the exact time because it was so intense, and we were working frantically, so no one had the time to look at the clock,' says Alberto, and I hear his voice thicken again.

After the little kid had died, they opened up the body and discovered that she had a malformation of the lungs. Instead of two lungs, there were three, and the third took up all the space available, preventing the others from filling with air. The third lung was hard and compact, like a piece of liver.

'We don't know why that happened, whether it was something that went wrong during the cloning, or whether there were natural reasons for the malformation. It's not an unusual foetal malformation in goats, but it may have had something to do with the cloning — I really don't know,' says Alberto.

He tries to explain how he set aside the sadness and disappointment he felt.

'Each of these steps took so much effort; you can't imagine how much work it was and how many problems we had. So when the first kid died, I just saw that as yet another aspect of the hard work facing us, and thought we'd have to work even harder next time.'

The first time I talk to Alberto, we stay on the phone for longer than either of us had actually intended, and, just when he is getting to this part, I hear something starting to ping and beep in his office.

'Oh — could you ring me back in a few minutes? There's something I have to deal with,' he says.

When I call him back, he tells me he has had to switch off the computer and a few other devices in his office because the electricity has gone off. This is because the Spanish authorities switch off the electricity in all publicly owned buildings at four o'clock on Friday afternoons, and don't put it on again until Monday morning, to save power and money. The same financial crisis hit the bucardo project; after 2003, the money dried up, and the authorities stopped funding the project.

'I can understand that — it's not really morally right to put money into a project like this when there are so many other needs in the country,' says Alberto.

Lack of money is an issue that crops up constantly among the scientists I talk to. It's easy to think that something as groundbreaking and awe-inspiring as de-extinction would attract large amounts of cash. When I first heard about it, I pictured spacious, well-equipped labs and hordes of committed scientists. In reality, while the commitment is there, the money isn't.

George Church is building his mammoth on the side, alongside his usual research and without very much funding for that specific project. Ben Novak's salary is paid by Stewart Brand's organisation, Revive & Restore, but he's the only scientist they employ. The main purpose of the lab where he works is to analyse ancient DNA, and the passenger pigeon is a side-project for them as well. The same applies to the scientists I meet later. None of these de-extinction projects are particularly flush with money.

And money is one of critics' commonest arguments against the various projects to resurrect extinct animals. They argue that there are so few resources available to protect the species living today that it's immoral to put money into reviving species we've already lost.

In the ongoing Spanish economic crisis, Celia's cells are still on ice, while Alberto began working on other projects some time ago. A year or so ago, the researchers received a little money, and they checked that the cells were still alive and could be developed into an embryo. But no further progress has been made with the project, and there have been no more pregnancies.

Moreover, in the winter of 2014–15, something else happened that further complicates the situation. Ibex belonging to another subspecies escaped from an enclosure on the French side of the Pyrenees and seem to have managed to survive in the wild. Earlier attempts to introduce other types of ibex into the cold and inhospitable mountains failed, but these animals appear to have survived an unusually harsh winter.

'All this time, I've been working towards the goal of bringing ibex back to the mountains. Everything pointed to that being quite impossible without the bucardo. But if these animals have survived, if they continue to survive, that'll mean a radical change,' says Alberto.

He thinks that even if the new ibex survive, they might benefit from bucardo genes. A clone of Celia might be able to have young with males of this subspecies; alternatively, artificial fertilisation using her cells and their sperm might work. That would give the animals characteristics that would make them more resilient in winter, Alberto argues.

After over 25 years' efforts to save the bucardo, Alberto now feels as if the new ibex have pulled the rug from under his feet. He has yet to digest the new development.

'I really don't know what to believe or to think now.'

Today, Celia stands in the visitors' centre in the national park where she lived and died, but Alberto hasn't been to see her.

'No, I don't want to go and look at her now she's dead. Maybe you can imagine how I feel?'

CHAPTER 6

The Rhino That Came in from the Cold

There's something about Nola that reminds me of my grand-mother, who was fond of loose, colourful kaftans and — so it seemed to me — felt at ease, in her element, wherever she went. Though Nola is rotund and rather cumbersome, her movements are dignified and majestic, if not exactly rapid. She has no real worries in life, each sunny new day resembling the one before. She loves apples and relishes a thorough back scratch. The elderly male rhino nearby is kept firmly in his place. It's clear who rules the roost in their enclosure.

People are fascinated by endings, boundaries, the last of a line. Maybe that explains why I feel more moved than usual as I watch the monumental rhinoceros taking a leisurely stroll down to her feeding place to see if anything tasty has appeared. When I visited her at San Diego Zoo in southern California, there were only five northern white rhinos (also known as northern square-lipped rhinos) left in the world. A few months later, just four remained. The other three live in Kenya: Najin, another elderly female; her father, Sudan, the only male; and her daughter, Fatu.

'I've sometimes heard the curators say that Nola doesn't run about any longer, but I've seen her charge like a steam locomotive when she's riled by her mate and determined to see him off,' says Darra Davis, the zoo education worker who's showing me around. Nola's gentleman friend is a rhino of a different species, the southern white rhino. The main reason for his presence is to provide some company; all hope that the two might produce offspring together has long since faded.

Although Nola continues to munch her apples, she belongs to a species that has died out. Like Najin, she has passed calving age. Though Fatu is only 16 years old, there seems to be something wrong with her reproductive system. No matter how much IVF treatment these rhinos are given or how often they are inseminated, no more northern white rhinos will be born by natural means.

In November 2015, while I was writing this chapter, the news arrived that Nola had died of a bacterial infection and old age. She had reached the age of 41, which is like a woman living into her 80s. Then, in March of 2018, as I was revising the English translation of this manuscipt, I was reached by the news that Sudan, the last male of the species, had passed away. So now there are only two northern white rhinos left in the world, and hopefully they are still alive as you read this.

It is mainly poaching that has killed the northern white rhinoceros, slowly but implacably. Rhino horn commands as high a price on the black market as cocaine or gold. Some horns are used in alternative medicine that has no effect; others are turned into the hafts of ornate knives or other decorative artefacts.

'This is basically about whether we believe we have a responsibility to future generations. Most people agree we have a responsibility to our children and grandchildren, and that maybe we aren't responsible for the generations that may live tens of thousands of years in the future. But the decisions we take today will have consequences for that sort of length of time,' says Oliver Ryder in his office, a few hundred metres away from Nola. A researcher, he's responsible for the zoo's genetics department.

Oliver is not just talking about the fact that species are dying out at an alarming rate today and that the Earth's biodiversity is steadily shrinking. He's also talking about how the decisions we take could make the world better and help save species. Oliver is very fond of rhinos, and has been ever since the first specimens arrived at the zoo when he was a doctoral student here. Since then he has seen one species, the southern white rhino, thrive and recover, while the other, the northern species to which Nola belongs, has disappeared. There is an ongoing scientific debate about whether they are actually different species, or, rather, subspecies of the same species. Whatever the answer, it's clear that the white rhinos which once lived in eastern and central Africa have now disappeared.

'We need to talk about how we actually define species extinction. The definition we have today — that a species becomes extinct when the last individual dies — doesn't really make much sense. In practice, species die out long before that. It happens when they lose their capacity to reproduce, or when genetic variation is so depleted that the species can't survive in the long term. We don't always know quite where the boundary lies.'

Although the northern white rhino is already extinct in practice, that doesn't mean all hope is lost. Here, in the zoo just outside San Diego, another, quite different kind of zoo exists, packed in plastic tubes and immersed in liquid nitrogen. Six large containers hold tens of thousands of small test tubes full of cells, eggs, sperm, and a few embryos from about a thousand species of animals. When Oliver opens one of the containers, cold vapour from the liquid nitrogen wafts out, and he needs the thick purple rubber gloves he has on to protect his hands from injury. Slowly, he lifts out the receptacle containing cells from 12 northern white rhinos.

Twelve unrelated individuals may be enough to enable the species to recover. Twelve test tubes could enable new baby rhinos to rumble about once more like miniature armoured vehicles.

The big difference between the cells lying frozen in the zoo's basement and the bodies conserved through taxidermy or storage in formalin in the world's museums is that these cells are still alive.

'It's the cells that contain life. Any attempt to revive an animal species must involve living cells. Many people are obsessed by genes. But you can't create life from DNA alone,' says Oliver, lowering the test tubes back into their receptacle and pulling off his gloves.

If some of these cells are thawed in a dish containing a nutrient solution, they will start to grow, divide, and multiply. This makes the cells a renewable resource and means they can be kept almost indefinitely, Oliver explains. No one knows exactly how long frozen cells remain viable, but the cells frozen by scientists in 1976, when the collection began, are still in excellent condition.

Each 'frozen individual' is represented by eight test tubes, and each test tube contains about ten million cells. In most cases, they're derived from biopsies taken from living creatures, though it's also possible to take cell samples from animals that have just died. Half are stored here, with the other half in central San Diego, a precaution against anything from power cuts to the fires that sometimes rage in dry California.

'Ninety-nine point nine per cent of the cells we have here in a frozen state come from individuals that are now dead, even if they were alive when we took the cell samples. These are incredibly valuable samples. This is something absolutely irreplaceable,' says Oliver.

Among the test tubes, there are species that have already become extinct. Oliver tells me about the Hawaiian poo-uli, a small grey bird with a black mask around its eyes. Scientists realised the species was threatened, and they discussed whether they should go into the forests to try and catch the last few individuals. Maybe, the scientists thought, the birds might be able to breed in captivity, so their young could be released into the wild at some point in the future. But these discussions went on for a long time. While the scientists were talking, the numbers of birds dwindled. By 2002, there were only three individuals left. In 2004, the last male was caught, but the scientists failed to find a female for him to mate with before his death a few months later. The body was sent to Oliver.

'It was around Christmas, and I was sitting at the microscope examining the cells when it really hit me — a sharp, intense realisation that this species was gone now. I think it really affected all of us who were working on it.'

Oliver has been in charge of the Frozen Zoo since 1986. This is by far the biggest bank of frozen animal cells anywhere, though there are similar projects in other parts of the world. The largest bank of frozen cells in existence stores not animal but plant cells: a former coalmine on the Norwegian island of Spitsbergen, in the Svalbard archipelago, contains seeds from over 4,000 different types of plants, and the facility has sufficient capacity to store four-and-a-half million seeds. The Svalbard Global Seed Vault has been dubbed the 'doomsday vault'. The six containers in the basement here seem rather insignificant by comparison, and the scale of the task that remains becomes clear.

The thousand species represented here are mainly mammals, though there are quite a few birds, reptiles, and amphibians, too. Impressive though this collection is, it represents no more than a tiny fraction of all the vertebrates in existence today. For instance, scientists put the number of mammal species at over 5,000, ranging from the rhinoceros all the way down to the tiniest bat, to say nothing of all the other types of animal in existence. As more species become threatened, the chances of filling in the gaps in the library of cells dwindle. One acute example is the African forest elephant, which, according to certain estimates, is at risk of extinction through poaching within just a decade.

'It breaks my heart to see pictures of animals that have been killed on the news. It's particularly bitter when you know that anyone who was there when the picture was taken could have collected samples for us — if only they'd had some basic training. Now the species may disappear without even leaving any frozen cells behind,' says Oliver.

Thanks to his work on the frozen cells, he lives with a constant awareness of the presence and finality of death.

'Our goal has never been to establish a mausoleum; we want to help preserve these species. I don't think anyone realised what potential there was in the cells when we started. But later on came the research that proved it's possible to produce living creatures from skin cells.'

That was one of the greatest scientific breakthroughs of the new millennium. In 2006, a Japanese researcher, Shinya Yamanaka, showed he could take ordinary skin cells from mice and transform them into stem cells. For a long time, scientists had believed the process of specialisation to be irrevocable, but

Yamanaka proved it could be reversed.

The fact that stem cells can be obtained in this way has aroused tremendous hope in medical research circles. In theory, such cells would make it possible to grow new organs from a patient's own cells, or to repair injury or damage. It would also be feasible to transform the skin cells in Oliver's cold store into new living creatures.

Jeanne Loring has a wonderful voice, deep and smoky, with a husky quality accentuated by the fact that she has a cold when I meet her. She is a professor of neurobiology and the head of the Reproductive Medicine Centre at the Scripps Research Institute, also based in San Diego. Her main occupation is medical research. One of her major research areas involves using stem cells to cure Parkinson's disease. Trying to save the northern white rhino in cooperation with Oliver is a side project.

'If I had a million dollars and had to choose between investing them in conserving habitat for threatened species, or in genetic research into those species, I'm assuming it's a lot more important to conserve habitat and protect animals against poaching. But the northern white rhino is a species that's going to die out, and we need to take a decision about whether we're going to try to rescue them, and to what lengths we would go to rescue them. There's no alternative,' she says.

She and Oliver have been friends a long time and have talked a great deal about how her medical research could help save the endangered species he has in cold storage. But no practical steps were taken until 2007. In that year, when Jeanne's lab moved into new premises, she celebrated the move by organising a safari in the San Diego Wild Animal Park for all the staff, together with a mini-conference on stem cells and endangered species.

They enjoyed themselves tremendously, and on their return one of the young scientists said she wanted to try to convert some of the skin cells in the Frozen Zoo into stem cells. There was absolutely no guarantee of success; if anything, the reverse was true. At the time, the only species in which the technique had been successful were mice and humans. This was just one year after the technique had first been used.

Oliver recommended they try the method on rhinos and the endangered drill monkey, which is restricted to a small area of West Africa. It was pretty complicated, says Jeanne, but finally they managed to transform ordinary cells from the endangered species, the rhino and the monkey, into stem cells. They used essentially the same technique that scientists use to convert ordinary human cells into stem cells, but had to adapt the process to the genetic make-up of these particular species.

Jeanne hopes that one day they will take all 12 of the lab's frozen rhino samples and apply the same process to them. The next logical step would be to try to get the stem cells to develop into embryos, thereby creating clones of rhinos that once lived. This is essentially what Alberto tried to do with the bucardo.

But Jeanne plans to take things a step further; she wants to apply a new method and transform the cells into sperm and unfertilised ova. If that were successful, it would be a major scientific breakthrough.

'There's simply no reason to think that you can't get something to work, just because it's never been done before.'

Several research teams are trying to develop ways of transforming stem cells into ova or sperm (gametes). If this succeeded it would, for instance, make it easier in some cases for childless people who want a family to have biological children of their own. Scientists have managed to trick cells into forming a preliminary stage on the way to gametes. When implanted in the testicles or ovaries of mice, these cells turned into sperm or ova. With human beings, too, scientists have managed to convert stem cells to this intermediary stage on the way to gametes.

Before new rhinos can be produced, the process needs to progress to the stage where the cells can develop into mature gametes in the lab. In the spring of 2016, a Chinese research team claimed that they had succeeded in doing precisely that, producing mouse sperm. At the time of writing, there's still some doubt about the result, as no other scientists have managed such a feat. Yet there's much to suggest that a breakthrough of this kind is on its way. Once the method has been developed — whether by Jeanne or by someone else — she wants to apply it to the frozen rhino cells.

Then the plan is to combine ova and sperm in every imaginable permutation, thereby maximising the genetic variation of the first generation of new rhinos. One advantage of this method is that cells from males and females alike can become either sperm or ova. This is the same method that has been discussed with a view to enabling gay couples to have children who are the biological product of both partners. In Jeanne's program, the resultant rhino embryos would be implanted in surrogate mothers, using standard IVF treatment. These mothers would be southern white rhino females, closely related to the northern white rhino but not endangered.

'The great thing about this technique is that we can also apply it to frozen cells from dead animals. We don't have to have living animals,' says Jeanne.

Producing new individuals from ova and sperm was not the plan to begin with, she tells me. She had discussed with Oliver whether her research could help living animals from endangered species suffering from diseases such as arthritis or a heart condition.

'But once we'd finally managed to produce such great stem cells, the only thing that really made sense was to go for the gametes.' Jeanne, too, has fallen under the spell of the rhino, and her office is full of little statues of the animal.

'I didn't really have a specific interest in rhinos from the start, but they're close to my heart now, and I've gotten very interested in them.'

Jeanne doesn't want to guess how long it could take for the first new northern white rhinoceros to be born. Oliver thinks the main thing is for their work to continue.

'Personally, my attitude is that we have to do everything we can to save the species living today. As for the northern white rhino, I think we should give it a try. We have the tools, and I don't like giving up and saying it's impossible.'

While I stand watching Nola eating in the enclosure, it seems like a no-brainer to agree. If it's in the scientists' power to save this species, I want it done. The northern white rhinoceros is already an extinct species in practice. Yet from another perspective, it is a species that will never disappear. As long as nothing happens to the containers full of liquid nitrogen, the cells from the 12 individuals will remain where they are, a promise for the future.

The big question that gradually takes shape in my head is whether this is an empty promise. I think about Ben and the passenger pigeons he's trying to create. Rhinos created from these cells will be far closer to their ancestors than Ben's pigeons would be. It is in the cells that life resides, and maybe one can say that since the cells are still alive, the animals have not died either.

Yet something has been lost. To claim otherwise would be absurd. It may be possible for new rhinos to replace the old ones, but they wouldn't be identical. On the other hand, the rhinos running around in Africa 50 years ago weren't the same as those that lived 100,000 years ago. Does that matter? The Frozen Zoo will be an attempt to stop the clock, to preserve nature as it is now. But why should nature as it is now be of any greater value than the natural world of 10,000 years ago, or the species that will exist 10,000 years from now?

Right now, I have no idea, but my stomach knots up at the thought of Nola's death, at the idea that everything that is unique in her and her species will cease to exist. The feeling that this should not happen is stronger than the philosophical quibbles in my mind.

When we continue our conversation in his office, it becomes clear that Oliver has even bigger plans for the possible applications of the frozen cells. In his eyes, their main potential is to be used in genetic rescue operations to support species currently on the brink of extinction.

A common problem with endangered species is this: so few individuals remain that there's a huge risk of inbreeding and genetic problems. Even once a species has begun to recover in numbers, these problems remain because it has squeezed through a bottleneck, with very few individuals, and lost much of its genetic diversity. It can take a long time to build up diversity again; during that time, the animals of that species are particularly vulnerable to disease and other types of damage.

'They sometimes talk these days about genetic rescue operations that involve moving individuals from one area to another, to bring in new blood. The wolves of Scandinavia are an example. But for certain species, the frozen cells are the only way of introducing diversity; they are the only other individuals in existence,' Oliver says.

Soon there will be a way for scientists to clone individuals from the frozen cells, or to transform them into ova or sperm that can be combined with those of existing animals. But there will also be a different way for genes to be used. Oliver talks about an example, the California condor. A large, black, vulture-like bird with a hard face and a large, hooked beak, it was in such danger of extinction that in 1987 scientists decided to catch all the remaining wild individuals — 22 of them — and start a breeding program in captivity. That was a success, and the birds have begun being released into the wild again. But today they are threatened by a genetic disease that affects the chicks inside the eggs, rendering them unable to crack the eggshell and emerge. Oliver's frozen samples from condors could enable scientists to research the disease and find remedies.

'We have samples from every specimen of the California condor alive today. Suddenly, we're in a whole new world of possible ways to affect the preservation of endangered species.'

Among the samples, there are frozen cells that could give a genetic boost to many different species, Oliver explains — from frogs and gorillas to black-footed ferrets. The different ways in which the cells can be used are increasing all the time, and will develop in ways he doesn't even dare predict.

'The aim is to conserve species and to use the new technology to do that better. It's us who are now designing the future for genetic diversity. It's not about making lists of species any longer. We can design life and nature as we want them, create species.

Would we like to have miniature rhinos as pets? Would we like to have domestic cats that look like tigers? If we have that, do we still want to have big tigers living in the wild? What do we want the world to look like?' he asks rhetorically.

But that question is no longer merely rhetorical. Genetic engineering to develop new, exotic pets is a development that has picked up speed over the last few years. There is a Chinese company that sells genetically modified miniature pigs, which grow to no more than 15 kilos. It's planning to let customers decide in advance exactly what patches of colour their pigs should have and then modify each piglet to meet the specification. The same company also wants to develop genetically modified koi carp with patches whose colour, shape, and patterning can be determined by the customer. Decorative carp are an industry worth millions in Asia, and the firm is expecting to earn a great deal of money if they start to sell their variants. An Australian researcher is examining ways of modifying dogs; others are looking at racing horses. I am convinced it's only a matter of time until Oliver's visions of the future are reality.

He's aware that the idea of using advanced, pioneering genetic techniques to recreate extinct species is controversial. When I ask whether he himself thinks it's a good idea to bring mammoths and passenger pigeons back to life, he doesn't really want to answer.

'We have to be able to answer the question of why we want to do this. That isn't a scientific question, it's an ethical one. Are we going to recreate extinct animals? How extinct? Animals that have just died out, or those that became extinct a few decades ago, or thousands of years ago? That's an ethical question, that's how we should formulate it, and we need to answer it collectively as a society.'

We are already manipulating nature in so many ways, says Oliver. He looks quite tired and down as we carry on talking about where these new techniques could lead. It's not only about how we kill off species, but also about how we shape nature. Like Stewart Brand, he takes the cod off the American coasts as an example. Cod are not extinct, they're not even endangered, but fishing has reduced both numbers and size. They can no longer fill the same ecological niche in nature that they used to fill. The same applies to many other species, both marine and land-based. There are no places left in the world that we humans haven't already reshaped and affected.

I rebel against this reasoning. Surely there must be some remaining wild places and natural surroundings that are untouched by human beings? Surely there must be limits to our impact on the planet? But Oliver is implacable.

'The notion that there is a Garden of Eden somewhere, nature unaffected by human beings, is no longer possible. That isn't how nature works any longer. In the end, it's us who design nature. Our influence is huge, and it's going to get even more dominant and invasive. It's going to happen in ways we can't even really imagine or understand today,' he says.

That sounds to me like a terrifying scenario, a dystopia as bad as George Orwell's *1984* or Suzanne Collins' *The Hunger Games*. It is not a world I want to live in.

But Oliver also sees potential in such a future. He says we humans need to start thinking about our role, about how we should exercise our stewardship of the natural world. If we were to do that, we could use our enormous influence to make the world better instead.

'Today we are a force that destroys biodiversity and brings

species to extinction. But we could become the first species in biological history to consciously expand biodiversity,' he says, and carries on outlining how larger, more extensive frozen zoos could contribute to that goal.

'We can do that if we save and look after frozen cells properly. We can look at this from a longer perspective. Evolution causes species to disappear and appear, often quite rapidly. We're currently in the midst of Earth's sixth mass extinction, and lots of species have disappeared because of us. But if we were to systematically start collecting cells, we'd be able to increase the number of species in the world instead, to conserve species that would otherwise have disappeared. Today, we're spreading destruction around us, but we could also consciously increase the number of species in future with the help of cell banks like this one.'

Again, he echoes Steward Brand in his vision — a world full of biological treasures, and people who are sensible and responsible enough to act as benevolent stewards. This dream reminds me of the peace-loving world of Captain Kirk, driven by the love of exploration. Though that's a tempting picture, something about it makes me uncomfortable. It looks altogether too easy.

CHAPTER 7

'It's Not Quite That Simple'

'Many of the biologists who're working to save species on the verge of extinction see all this as high-tech nonsense; they think it's a waste of time to talk about de-extinction when we can't even manage to save the species we have now. Personally, I think that's a simplistic argument.'

Phil Seddon combines a light New Zealand accent with a dry British humour that shines through in the course of our conversation. His specialism is introduction biology: trying to save species by moving them to new locations, or releasing animals from zoos so they can return to the wild. In parallel with his research, he works for the IUCN (International Union for Conservation of Nature), developing guidelines for scientists who may release revived species into the wild in future.

'I think it'll happen. Those of us whose work involves species conservation have to have our say in this debate. There's a small window of opportunity now for us to have a serious discussion about how to use this technology. Now's the time to talk about how we can maximise the advantages and minimise the risks this option presents.'

All the scientists I've spoken to who are involved in de-extinction want the species they revive to be released back into the wild. George Church doesn't want to build a mammoth just to show that he can; he wants to see a hundred thousand mammoths roaming free in the Siberian tundra. Ben Novak wants to release his passenger pigeons into the forests. Speaking for the movement, Stewart Brand describes it as an opportunity for humans to repair some of the damage we have done to the world.

However, the response from biologists working to save animals on the verge of extinction has not been as enthusiastic as Stewart may have hoped.

'I don't actually know for sure, but I'd guess it probably came as quite a shock to the geneticists — the ones working to develop these technologies — that conservation biologists weren't all overjoyed at the idea of de-extinction. You'd think it'd be a dream come true for us, after all. But it's not that simple,' says Phil.

The reason is that so much can go wrong when animals are released into the wild and left to fend for themselves. The first and most obvious problem is that they may be simply unable to cope. Examples of scientists failing to translocate species successfully are legion. Sometimes the problem is that the animals lack the knowledge they need for survival; they may, for instance, be unable to hunt or protect themselves against predators. If a furry little mammoth is ever born in George Church's lab, there will be no mother mammoth, no one to show the calf how to dig up frozen grass covered by a thick layer of snow, which way to turn in a snowstorm, or how to rid itself of the maddening mosquitoes that land on its trunk. The calf will have to work out all these things for itself, along with all the hundreds of other skills that once constituted mammoth lore, and which young mammoths learned from their parents.

This is why Ben Novak has come up with his ingenious plan for introducing carrier pigeons that can coax new passenger pigeons into flight. It's easy to forget, but much animal behaviour is based on learned skills rather than instinct. Animals born in labs are like any others in that they have to learn what to do in order to survive. The same applies to those that grow up in zoos; though they may have their parents nearby, life in an enclosure is totally different from life in the wild. Animals that thrive in captivity aren't necessarily successful in the wild. In practice, captive animals always become slightly tamer, with all the changes that implies. Wolves, for instance, take the first step towards becoming dogs. For some species, that makes it far harder to revert to wolfdom on release.

Despite all the difficulties, there are also plenty of success stories — animals that biologists have helped to survive and thrive in the wild again. Expertise in this area is growing all the time. For instance, biologists can raise fledglings using puppets modelled on adult birds, which avoids over-familiarising the young birds with humans. In Sweden, to give another example, the peregrine falcon has been saved by raising birds in captivity and releasing them into the wild.

Some species have even been saved this way when all hope seemed lost.

Phil Seddon tells me the story of the Chatham Island robin, a small black bird that lived on an island off New Zealand. By 1980, there were only five birds left. Scientists decided to catch them to protect the species, but discovered that there was only one female able to lay eggs. They arranged for other birds to incubate her clutch so she could lay more, and, little by little, fledglings hatched, grew to adulthood, and were able to raise offspring of their own. The young black robins were released back into the wild, and today there are over 250 of them on Chatham Island. In fact, the species is no longer classed as seriously endangered. Moreover, although the birds are all descended from a single hen robin, they seem to have avoided any serious genetic problems.

However, if a resurrected species such as the passenger pigeon were to thrive in the forests of the north-eastern United States and recover its old habit of gathering in dense flocks, that wouldn't mean all the possible problems were solved. In fact, the reverse is true. After all, there is a major risk that the revived species itself could cause difficulties. Australia's various invasive species are probably the best-known examples of negative consequences. So how should we approach the risks and opportunities involved?

I ask Phil to give me a few examples of situations where he thinks it would be a bad idea to attempt to revive an extinct species, and others in which it would be positive.

'The default position should be: it's always a bad idea. We need to ask a few simple questions, I think. Firstly, do we really need to do this? Secondly, can we find an existing species that could be an ecological substitute? You shouldn't start thinking about the option of de-extinction until you've checked that the species is really necessary and no other species can replace it. You have to understand the risks and the uncertainty involved first of all.'

The most important thing to bear in mind, says Phil, is that the resurrected species will not be exact copies of the extinct ones. Not even clones will be identical to the original, as the way they've grown up will have shaped their behaviour. The differences will be even greater with genetically engineered species such as mammoths and passenger pigeons; these will inherit most of their characteristics from the original species, the elephant and the band-tailed pigeon. Their behaviour will depend on a combination of the new genes, their original genetic material, and the environment in which they grow to adulthood. All this makes it much harder to predict the outcome when these animals are released into the wild.

'In a project like this, you can never be quite sure how the creature you're recreating is going to function in the wild. Many biologists working with invasive species hate the idea. They've seen just how many problems new species can cause.'

There's one more aspect that makes resurrected species both more interesting and more problematic when it comes to releasing them. Many of the scientists seeking to create new versions of extinct species aren't aiming simply to bring back those particular flora or fauna; what they want above all is for the new species to have an impact on the environment into which they are released, to shape their surroundings. The passenger pigeon that Ben is trying to create is a case in point.

But this is a difficult thing to achieve, and the results are unpredictable, Phil says. He also thinks there are serious risks in romanticising the natural phenomena we have lost, in viewing them through the rose-tinted spectacles of nostalgia.

'Projects that aim to recreate a vanished ecosystem in one way or another are often based on the assumption that there was once a time when everything was wonderful, in perfect balance. There's a sense that if only we could recreate that state of affairs, everything would be just fine. But all environments are subject to constant change; nature has never stood still.'

Replacing one extinct species with another species is something biologists already do on occasion. It's an option if the extinct species was so important that its loss threatens a local ecosystem with collapse, and there is already a species available that could fill its niche.

One example is the giant tortoise native to islands in the Indian Ocean. Just about every island once had its own local species of giant tortoise, which lived in isolation from other similar species. They spent their days grazing, producing a particular kind of turf with closely cropped grass, known as 'tortoise turf'. But the tortoises vanished from a number of islands, generally because of humans. Their capacity to survive for a long time without food or water made them popular prey for sailors. Live tortoises were piled up in the ship's hold and eaten during the voyage. The islands were also invaded by proliferating stowaway rats, which devoured the tortoises' eggs.

When the tortoises disappeared, the turf also began to disappear, along with all the other species that depended on it. Weeds displaced the indigenous flora. This happened time and again, on island after island. So now scientists have translocated tortoises belonging to the surviving species to islands whose own species have died out. This has enabled tortoise turf to make a comeback.

'Maybe it doesn't sound like such a big deal to substitute one tortoise for another that's nearly identical. But translocation of species is a very sensitive subject for biologists — precisely because things have so often gone wrong in the past,' says Phil. Moreover, active conservation biologists tend to feel that the less we interfere with nature, the better.

Only once we have found a solution to the ongoing crisis will the time be ripe to think seriously about engineering substitutes for vanished species, Phil argues. If we are going to do such a thing, we need to weigh up the risks carefully.

'I don't know of any situation where it wouldn't be preferable to try and work with the species we already have,' he summarises.

The main philosophical problem with attempting to revive extinct species is that we may be fooling ourselves. We imagine these methods will enable us to recover something we've lost.

'No one would say that freezing the DNA of humans preserves what makes us human,' says journalist Maura O'Connor in an interview with *Wired*. She has written a book about attempts to save today's endangered species, and criticises the growing focus on preserving genetic material and freezing cells, when what we should be doing is conserving the natural environment and ecosystems in which endangered species can live.

If you were cloned, and your clone grew up in a different family, a different environment, and a different time, he or she would be a different person. A species that's restored to an ecosystem will be a piece in a brand-new puzzle. It will co-exist with other species that have evolved in its absence. It may be able to fulfil its old role, it may find a new one, it may not fit in at all — but ultimately, as Heraclitus said, you can't step into the same river twice. The world and nature are ever-changing.

In focusing on individual species, individual genes, we risk failing to see the wood for the trees, missing the puzzle itself through our myopic obsession with a single piece of it. Moreover, this focus risks fostering the illusion that white-coated scientists with sophisticated devices will intervene and save the day, enabling us to shelve all those awkward problems to do with people, politics, finance, and so on — everything that goes into ensuring that resurrected species actually have a habitat where they can live. Scientists are seen almost as comic-book superheroes, unencumbered by banal reality.

As I mentioned earlier, George Church sees his work not just as a way to restore the mammoth, but equally as a way of giving a new lease of life to Asian elephants. With a few additional features, they will be equipped to live in a new habitat, where they will no longer be threatened by hunting or deforestation. However, in Phil's view, going down that path would be tantamount to abandoning the possibility of conserving elephants where they live today. If their existing habitats disappear, many other species will be lost at the same time.

However, that doesn't mean these projects should be shut down, or that geneticists should stop trying to build mammoths, he says in the next breath. He's acutely and painfully aware that current efforts to rescue endangered species face serious difficulties, and often fail. The other side of the coin is the fact that today's biologists are not coping with the ongoing crisis.

'Maybe not too many people know this, but we're actually losing the battle.'

Phil explains that the method used to save both individual species and ecosystems has long been to change as little as possible, preserve what remains, and carefully try to restore the way it used to look. But that's no longer working. The situation is so drastic, and so much of the world is changing so radically, that traditional models such as national parks and protected areas are no longer enough. The focus on preventing individual species from dying out is not proving very effective. There are only a few successes among a great many failures.

'We're fighting fires all the time. No one has had the time or headspace to take a step back and say, "What are we trying to do? What is it we want to achieve?"'

He hopes all the interest in species revival will help boost commitment to conserving species that are currently under threat.

'We've been trying to hammer home the same old message for so long now: There's a crisis! We're losing species! And so on and so forth. I think the public are getting fed-up with hearing that message. Maybe they're thinking: "Oh God, haven't you solved that problem yet?" We have to find new ways of engaging people. Giving them bad news all the time hasn't worked terribly well. Maybe these attempts at de-extinction will give people new hope.'

The really positive thing that could come out of the de-extinction debate is that conservation biologists could be forced to give more thought to their methods and what they're actually aiming to achieve, while geneticists could make it clear that there are a great many new tools that could be used in conservation.

'I think this gives plenty of food for thought to those of us working in this area — or it should do, at any rate,' says Phil, with a mildly ironic laugh. 'I don't think it's something we can afford to ignore just because it sounds like a techno-fix. It's going to be very important one way or another, I think.'

In New Orleans, female cats run around in a courtyard outside a lab. Now and then, a white-coated scientist wearing blue surgical gloves picks one up and lays her on an operating table. Once she's been anaesthetised, her ova are carefully removed from her ovaries with a suction needle. While the cat reawakens, the ova are transferred to a special lab where the cell nuclei will be removed and replaced by different ones. They will grow into clones.

The ACRES lab specialises in cloning various kinds of cats. They use ordinary domestic cats as surrogate mothers and egg donors for rarer species under threat. The method used involves removing a cell nucleus containing genetic material from a rare cat and inserting it into an ovum from a domestic cat. The embryo that develops from this ovum is then implanted in a female domestic cat, who thus becomes the surrogate mother of the rare-cat clone. The lab's achievements include cloning an African wildcat, and demonstrating that a pair of cloned cats can have viable offspring together.

Bridging the gap between closely related species makes cloning even trickier. That was the challenge facing Alberto Fernández-Arias while he was working to clone Celia the bucardo. A number of other threatened animal species have been cloned by scientists in different parts of the world, but so far these efforts haven't been successful enough for the method to be used to produce animals that can later be released. The wildcat and the ordinary domestic cat are particularly close relatives, which is why the technique works well in this case, but scientists are working to develop the method so it can be applied to other species of cats. They've tried out the same process on the black-footed cat of southern Africa, but have not yet succeeded in cloning it.

Martha Gomez, the geneticist who heads this project, has said in several interviews that the aim is to help protect and conserve threatened species of cats. This can be achieved both by cloning them and by implanting embryos produced through IVF in domestic cats, which serve as surrogate mothers. This is something the ACRES lab has succeeded in doing with several species of cats.

So far, none of the cats born in the lab have been released. Much still needs to be done before that will be feasible. The lab's work still focuses on developing the technology itself. Although scientists have been cloning animals for 20 years now, they're still not very good at it. A few individuals belonging to endangered species have been cloned, including the mouflon (wild sheep) and the gaur and banteng (two types of bovine). But these efforts have been very resource-intensive, and many clones have died prematurely. For Martha Gomez and her staff, only one egg in 40 develops into a viable new individual. No one yet knows quite why the process is so difficult or what part of the cells causes these problems.

Given the many problems around cloning, some scientists question whether it can really play a significant role in protecting species on the brink of extinction. Scientists working in the field naturally hope to find ways to improve the technique and reduce these problems. For some animals, such as the northern white rhino, the new technologies represent the only possible way to save the species.

'I don't know when it's time to give up your hopes for a species; maybe you never give up, but look for any solution you can find. And here we have a technology that can help us and enable us to save species that we would once have lost,' says Phil, when I ask him what he thinks about the chances of cloning threatened species. His answer reminds me a little of saving premature newborns, the chances of which are constantly improving.

Yet as I've seen, biologists are divided on the issue. Many scientists see attempts to apply gene technology to saving or resurrecting species as a waste of time and resources. The concept of de-extinction has been roundly criticised by scientists, and feelings on the subject run high. The same criticism can be heard again and again: if we think species revival is so easy, why bother to protect endangered animals?

There's one thing that needs to be remembered in this debate: essentially, we all want the same thing. Everyone discussing the issue, from geneticists like George Church to visionaries like Stuart Brand and conservation biologists like Phil Seddon, is working towards the same goal — to minimise species extinction and maximise the world's biological diversity. All are deeply committed to improving the natural environment; all are tree huggers.

So the issue isn't whether species should be protected, or whether protecting them should be a priority. The question is how best to achieve that aim. The more I read all the criticisms made, and follow the cut and thrust of discussion between scientists, the more I come to view the whole thing as a kind of culture clash.

The geneticists, equipped with their new technologies, have made a rather flat-footed entrance in a research field of which they have quite limited knowledge. They have blithely demonstrated their methods, announcing: *Just look at this — it could be the solution to all your problems!* The conservation biologists, for their part, have been 'caught napping' to some extent. Since they haven't been following developments in gene technology, they fail to see all it has to offer. In their view, the geneticists are offering quick fixes to complex problems, fixes that are populist and unrealistic. They may also feel that the geneticists lack respect for the real difficulties involved in conserving and protecting species. Speaking as an outsider, I'd say they should all sit down in the same room and listen to each other properly over a cup of tea.

When I met Jeanne Loring, the scientist who transformed skin cells from the northern white rhino into stem cells, I asked her how conservation biologists reacted to the project.

'They're nervous. I can understand that. Money's tight in this area of research, and they're focused on conserving the natural environment. Possible high-tech solutions like this make them uncomfortable, and I don't think they really trust us yet. It's taken them a long time to get used to the idea.'

Phil Seddon's analysis is more or less the same as mine. The problem, as he sees it, is the lack of contact between the people developing sophisticated gene technology to revive lost species and those who are working to protect species in the wild. Neither really understands what the other group is doing or the kind of problems facing them. But that is changing.

'Steward Brand and Ryan Phelan of Revive & Restore have just held a conference for exactly this purpose. They brought the two groups together and got them to discuss the issues. I think most of the attendees found it very inspiring. For those of us who work in species conservation, this should lead to some really exciting discussions and new ideas.'

This is about using gene technology to develop brand-new tools for protecting endangered animals. Genetic engineering may prove decisive, turning the tide and making a difference in the battle to conserve species. One possible application of the new technology that's currently being developed is to support species that have gone through genetic bottlenecks, becoming highly inbred and vulnerable to genetic defects. This is something Oliver Ryder talked about when I met him in San Diego.

'The idea of boosting genetic diversity could be a tremendously exciting new track,' says Phil.

This implies using the frozen cells kept by Oliver Ryder and others, but possibly also drawing on other sources of genetic diversity. Perhaps the genetic make-up of some museum specimens could be analysed and used to understand and enhance species living today. This wouldn't produce enough individuals to start a new population, but it might lessen the risks facing endangered animals if they could benefit from an injection of new genetic material, a small shot of fresh blood into the system.

'We can't afford to ignore this development just because it sounds like a high-tech distraction. I think what's happening now will be hugely significant for species conservation, one way or another,' Phil continues.

Developments over the last five to ten years have proceeded at such a rate as to astound even scientists working in the field. The knowledge of what can be achieved hasn't yet found its way beyond the doors of the science labs and out into nature. But Phil is sure it will get there in the end.

'This is a development that's both exciting and worrying at the same time. We live in interesting times.'

CHAPTER 8
God's Toolkit

The temperature on Earth continues to rise inexorably as we humans release more and more greenhouse gas into the atmosphere. The growing quantity of carbon dioxide in the atmosphere is both causing the oceans to warm up and making them more acidic, as CO_2 is absorbed by water, forming carbonic acid. This is catastrophic for the world's coral reefs, which can't cope with either the temperature or the acidity. Their skeleton, composed of calcium, simply dissolves.

A possible solution may emerge in one of the huge tanks of seawater at the newly built Townsville Sea Simulator (SeaSim) in the north-eastern corner of Australia. These large tanks are being used to grow quantities of tiny coral fragments, under the watchful eyes of Madeleine van Oppen and her fellow scientists. Off the coast lies the Great Barrier Reef, home to over 600 species of coral — yet half of its species have disappeared over the last 30 years.

At the SeaSim, Madeleine is trying to nudge evolution so as to produce types of coral that can survive in the oceans of the future. One method is to cross different species of corals in the hope that one of the hybrids will have more robust characteristics than its predecessors. Another possibility is to alter the algae that live in symbiosis with the corals. Algae and coral pool their resources at a very early stage of the coral's development, becoming inseparable partners. If the algae die, the corals are doomed as well. Unfortunately, algae seem to be more sensitive than coral to warmer water. By having the same types of coral form a symbiotic relationship with many different types of algae, scientists are hoping to find a combination that is less sensitive to temperature. A third method is to allow various corals to grow in warm, acidic water, to see whether they adapt spontaneously in a way that might help them cope better.

The researchers hope to give the corals a helping hand in the world we humans are in the process of creating. Planting out corals that are better adapted to the new conditions may provide a possible way to save the world's coral reefs.

This is a controversial technique born of scientists' sense of desperation. Now that humans are changing the planet at such a rate that flora and fauna can scarcely keep pace, more and more scientists are thinking about possible ways to accelerate evolution. So far, Madeleine and her colleagues plan only to try to identify spontaneously occurring positive adaptations. But in future, they may conceivably have recourse to genetic engineering.

Today's scientists have the capacity to make far more exact and precise changes in both flora and fauna than in the past. Most scientific projects of this nature have to do with medical research. Examples include producing stem cells that can cure certain types of blindness, genetically modifying pigs so that their hearts can be transplanted into humans, and seeking ways to adapt human embryos or young children to save them from serious congenital diseases. Medical science has just begun to conduct the very first trials of genetic engineering as a means of treating patients.

In the autumn of 2015, Layla, a one-year-old girl with leukaemia, was injected with immune cells genetically modified to target her specific cancer. This was intended as a way to keep Layla alive until a bone marrow donor could be found, but has since demonstrated a high remission rate and wider use has been approved.

Layla's was only the second trial involving the use of gene-edited cells to treat patients. A year previously, gene technology had been trialled on 12 HIV patients. A number of similar trials involving a variety of medical conditions, including HIV and several types of cancer, are ongoing.

Stewart Brand tells me he sees de-extinction as an important factor in boosting cooperation between geneticists and conservation biologists. His goal is to make geneticists and biotechnology developers aware of species-conservation issues, so they consider possible ways of applying the technology in that context. As matters stand, there is a gulf between the two fields, and it's taking a long time for the new technology to trickle out and reach people working in species conservation.

'Biotechnology is driven by medical applications, but we want to start applying it as soon as possible. Our goal is to ensure that species-conservation technology is up to the mark. We shouldn't have the 20-year time-lag we've had so far,' he says. Combining the new gene technology with species conservation has become a key issue for Revive & Restore.

'That's one of the reasons why we're pushing for species revival at quite an early stage in the development of this technology. We want it to become a part of how this technology sees itself, one of the ways it can be used. Species conservation needs to be permanently on the agenda for the people developing gene technology.'

There are plenty of indications that he will succeed in making his vision come true. Phil Seddon may be critical of species revival as a concept, but he's enthusiastic and hopeful about applying gene technology to save the species alive today.

'The new genetic engineering technologies will change everything. Helping threatened species will be one of the applications of gene technology,' he says.

Phil's current focus is on possible ways of using gene technology to get rid of invasive mice on small Pacific islands near New Zealand. The mice cause huge problems, displacing birds and other small animals living on the islands. One of the ways that has been used to reduce their numbers is putting down poison, but there are several problems with that. The main risk is that other animals may ingest the poison; another is that the mice may learn to avoid it. That would mean they would not be completely eradicated — the mouse population would recover as soon as the scientists stopped putting down poison. Projects like this have also been criticised for causing unnecessary suffering to mice.

There are a few examples of scientists who've been completely successful in eradicating invasive species from islands, but this is difficult and often demands huge efforts. Now geneticists seem to have found a miraculous solution to the problem. The idea is to insert a kind of self-destructive gene into a number of mice that are then released on an island. As a result of this gene, the offspring of these mice would all have male progeny, but no females. In nature, such a mutation would quickly disappear through natural selection. But this is where a really sophisticated aspect of genetic engineering comes in.

The scientists would link this mutation to what is known as

a gene drive, a gene which ensures that all offspring — not just half of them, as is the norm — inherit a particular feature from a parent. As a result, that characteristic can spread much faster and isn't eliminated through the normal evolutionary process.

The effect on the mice living on the small island would be that after a few generations they would all be male and would therefore die out automatically. Birds and other small animals would recover, species would be saved, and unique ecosystems would be protected.

'I'd really like to see this tried, and there are plenty of islands where it could be tested. The technology is still very new, and there are lots questions that still need to be answered. But the people I've talked to are sure it would work,' says Phil.

Research into gene drives is a brand-new field. Only a handful of scientific studies have been published, based on laboratory testing, but the results are both promising and convincing. They are also frightening. There's so much that could go wrong.

If mice bearing a gene drive were to swim from an island to the mainland, the result could be the extinction of all mice. If a mouse with a gene drive mated with another type of mouse and they produced offspring, another species might become extinct. If both those things happened at the same time, it's hard to even imagine the possible consequences. Living in a world without mice might seem tempting to anyone who has to keep buying mouse traps for their attic, but the impact on nature would be huge.

Another possible application would be to use this technology to purposely eradicate a species. Some scientists advocate releasing gene drives among the mosquitoes that spread malaria or other diseases. They have even successfully trialled the genetic engineering of malarial mosquitoes in the laboratory. In that instance, the gene drive was linked to a mutation that makes all female mosquitoes sterile. If mosquitoes with that gene were released, the species might be eradicated in just a few years.

Another way of applying this technology would be to implant a gene that prevents mosquitoes from spreading disease. This technology has also been tested on malarial mosquitoes. It would have less of an impact on the ecosystem, as mosquitoes would continue to exist, but it would also mean that the diseases concerned might develop resistance and make a comeback. In another example, scientists are looking into the possibility of implanting gene drives in ticks that would stop them spreading Lyme disease.

Although the gene drive method is still new, it's the subject of vigorous debate by the world's geneticists precisely because the impact — both positive and negative — could be so huge. In light of this, scientists are trying to design safety measures to shut down the gene drive if anything went wrong, and the first such fail-safe systems are currently being trialled under laboratory conditions. Yet there's much to suggest that gene drives would be essentially irreversible if released into the genes of wild animals.

Trying to get rid of unwelcome animals and infections is one thing, but gene technology can also be used to try to protect wild animals against disease.

The mammoth is just one element of all that George Church aims to achieve. First, he wants to use the technology to protect today's Asian elephants against a virus, a variant of herpes that often kills young elephant calves. This virus is one of the main threats other than human beings currently facing Asian elephants.

'If we could eliminate the herpes virus, or if we could make sure it was no longer deadly to Asian elephants, that alone would reduce the risk of extinction. So even before we make an elephant that's resistant to cold, we could make one that's resistant to herpes,' he enthuses.

The aim is to find a cure: a vaccine, a means of engineering genetic antiviral treatments, or perhaps both. Phil Seddon also thinks gene technology could be very effective in combating diseases. He draws attention to the possibility of making frogs or bats resistant to the fungal diseases that currently pose a threat to these two groups of fauna.

'If we could make frogs resistant to fungal infections, for example, we'd have an excellent reason to do so.'

As a result of the current global epidemic among frogs, a third of all species are threatened by extinction. The fungus *Batrachochytrium dendrobatidis* is the chief problem. It has spread over the globe like wildfire, and can wipe out entire species in a flash. The fungus grows on frogs' skin, making it thick and hard. As a result, the affected frogs can't absorb moisture or other nourishment through the skin, so they soon die.

The world's bats are currently threatened by a similar disease, white-nose syndrome. A fungus that grows on bats' noses and wings has killed over five-and-a-half million of them in the United States. Scientists believe both diseases could be stopped by applying gene technology, and a range of possible solutions are currently under review.

As I read about the various projects and talk to researchers, it becomes clear that the putative applications range over a broad spectrum. To advocate the eradication of a disease that threatens the world's frogs is quite a straightforward matter. Eradicating the mosquito that spreads malaria is a weightier ethical issue. However, the thought that over 400,000 human lives a year could be saved makes it clear to me that it would be a good thing. You can hardly object, provided that the technology develops further and becomes effective enough.

When you start to think about altering corals and other species to enable them to cope with climate change, it becomes trickier. There are various ways of achieving this, some of which involve sophisticated genetic engineering, while others do not. For instance, scientists have recently identified a natural mutation in a species of salmon that enables them to survive in warmer water, and this mutant gene could be transplanted into other salmon.

Wouldn't it be better to do what we can to prevent global warming, instead of modifying nature? The answer is: yes, of course it would. But unfortunately, real life isn't quite that straightforward. Since humanity seems completely incapable of restricting temperature rises to within reasonable limits, it may be better for the world's scientists to do their best to mitigate the impact.

But what happens next? This is where things get really tricky. Should we manipulate species to enable them to tolerate other phenomena caused by humans? In theory, we'll very soon have the option of shaping nature to be just the way we want it to be. It's difficult to know where the changes will end.

'We're starting to ask some pretty hard questions about where we'll end up. I think this is an exciting time right now; it's exciting that we can discuss these issues,' says Phil.

Of course, not everyone takes a positive view of such change. People can be divided into two distinct groups. One comprises those who say the situation today is so serious that we must take this kind of action to have any chance of saving some of the biodiversity we still have. The other group is made up of those who say that any such change would mean abandoning the goal of conserving nature as it is, that nature modified by human beings would be something entirely different, representing an utterly different world. Moreover, they argue, there would be incredible risks if anything went wrong and got out of our control.

'The thing is, we can artificially create wild species that are better adapted to a world affected and modified by human beings. We can make them more resistant to lead, more resistant to pesticides, more heat- or drought-resistant,' says George Church. He thinks this is a positive development and we should pursue it.

'This technology doesn't change species in essence; it just makes them better adapted to today's environment. That's just like all the mutations that have enabled people to adapt to urban life over the last few millennia. They haven't stopped us being human,' he continues. His enthusiasm about the new possibilities that would open up shines out as he speaks.

I feel very uncomfortable. It is difficult, even in theory, to draw a line between what I think is right, and something that just seems terrifying. The new technology will make it possible to create brand-new species, to modify animals in ways that would otherwise be inconceivable.

'I think it's precisely this aspect that opens up great new opportunities for conservation biology. This is about making new species that are better adapted to the modern environment — which is actually much better than making new versions of old ones, almost as an obsession,' says George.

I ask him if he worries at all about the various ways in which this kind of technology might go awry.

'Of course I do — I'm a real worrier, I worry about everything imaginable. And I mean everything. I worry about what might happen if we don't do this, too. But I think the best way to deal with it is to plan, and to picture as many different scenarios as possible, then carry out a few small-scale tests to check our hypotheses, just like you do before putting a new drug on the market.'

But what limits can George see from the vantage point of his Harvard lab?

'I don't specialise in limits — my speciality is overcoming them,' he says with a chuckle. He looks more like Father Christmas than ever, but I know he's utterly serious.

CHAPTER 9

The Growing Dead

Death came to the United States in 1878, in the guise of a cargo of chestnuts from Japan. The shipment included boxes of sweet chestnuts and shoots from beautiful Asian trees. Smaller than its American cousin, the Japanese chestnut tree is cultivated both for its ornamental qualities and for its edible nuts. Samuel B. Parsons Jr, the landscape architect who took delivery of the shipment in Manhattan, began selling saplings to orchards all over the country. Unbeknown to him, he was selling another species together with the trees — the *Cryphonectria parasitica* fungus.

Japanese chestnuts are resistant to this fungus, which lives under their bark. Since the blight is undetectable from the outside, Parsons' mistake was easily made — but it was catastrophic for the native American species. In the eastern half of the United States, from Mississippi in the south to Maine in the north, American chestnuts were so abundant that they accounted for a quarter of all deciduous forest trees. There are descriptions of hillsides that appeared to be blanketed in snow when the trees were decked in the white blossom of spring. Chestnuts provided food for everything from squirrels, passenger pigeons, and insects to people. They were considered tastier than the European variety, and were ground into flour for cakes, roasted over open fires, candied, or used in brewing. The timber was used in house building and the bark for tanning leather.

American chestnuts were stately trees that could grow to 30 metres and live for over 100 years, yet they were defenceless against the new blight. After colonising a space between the bark and the wood of the trunk, the parasite exudes an acid that kills the tree's tissues, allowing the fungus to feed on the dead remains. The dead tissue forms a kind of canker, which spreads over the trunk, blocking the transport of nutrients and water between the roots and the leaves. The effect is the same as killing a tree by ringbarking.

The disease spread like wildfire. Within a mere 50 years, it had killed three billion chestnut trees, almost all the chestnuts in the American forests. A few great, majestic chestnuts remain, but nearly all of them were deliberately planted, and they are a long way from the areas to which American chestnuts were once native. They can be found, for example, in the west-coast states of California and Washington. However, the root systems of some of the affected trees still survive in the great forests of the eastern United States. They are like zombies, constantly reawakening and producing the odd sapling that manages to grow for a few years, until the blight takes hold and kills it again. The parasite can survive in the bark of various other species of trees without harming them, so the eastern forests will never be blight-free.

In practice, this means the American chestnut is an extinct species. There may be survivors, but the forests have lost everything the chestnut trees once provided, from pollen and nectar to insects in spring and the bountiful harvest of sweet chestnuts in autumn. The American landscape changed dramatically when these trees disappeared, and today the forests are full of other species.

The airfield at Chersky is a flat gravelled area beside the river. Only small propeller planes land here. The town, where many houses now stand empty, can be seen in the background.

Researcher Sergey Zimov, who has run the research station here since the 1980s. It was the location's freedom from Soviet propaganda and his love of the natural world that kept him here.

Just like other elephants, mammoths had only four teeth in their jaw at any one time. A well-preserved whole tooth such as the one shown here can weigh nearly two kilograms.

The research station run by the Zimov family, housed in a former TV station.

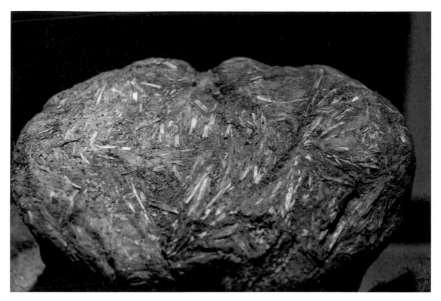

The permafrost has preserved not just mammoths' bodies, but their dung as well. This fibrous lump of excrement is displayed at the Mammoth Museum in Yakutsk.

The walls of the long tunnels in the Permafrost Kingdom are covered in hoarfrost. There is colourful lighting and tinkling piano music everywhere. The passageways are full of ice sculptures and the bones of mammoths and other creatures.

Among all the coloured lights in the Permafrost Kingdom, you can also find an awe-inspiring model of a mammoth, with real tusks.

Vestiges of hide and hair are still visible on this well-preserved skull. There are blocks of ice piled up along the walls to keep the temperature low, despite the tourists.

This Petri dish contains the basic material for the mammoth that George Church wants to build: cells from elephant hides, modified with mammoth genes.

George Church, based at Harvard, is an extremely optimistic and enthusiastic person. Trying to resurrect the mammoth is only one of his many projects.

Stewart Brand, together with his wife, Ryan Phelan, set up the organisation Revive & Restore, which has brought together the various projects aiming to recreate extinct animals.

Martha, the very last passenger pigeon, died in 1914. Now her preserved body can be seen at the Smithsonian Institute in Washington, DC.

There are hundreds of specimens of passenger pigeons in museums worldwide. This brightly coloured male can be seen in Lund University's zoological collections.

Male passenger pigeons had brighter plumage than females. Recreating this difference between the sexes will present a challenge to scientists.

Researcher Ben Novak shows a set of test tubes containing genetic material from a number of passenger-pigeon specimens. Each tiny test tube contains DNA from one bird.

The genetic material in stuffed birds breaks down rapidly, making it hard for scientists to analyse it. It is preserved best in the fleshiest parts of the bird's toes.

Nola was one of the last four northern white rhinos when I saw her in 2015. She died later in the same year. Following the death of the male Sudan in 2018, there are now only two surviving individuals.

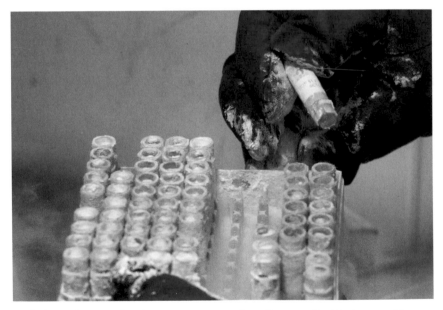

Cells from threatened and extinct species are kept frozen in liquid nitrogen. These cells are still alive and will continue to divide if the samples are removed from cold storage and thawed.

Oliver Ryder stands next to one of the large tanks full of frozen cells taken from over a thousand different species.

Jeanne Loring has successfully converted frozen cells from white rhinos into stem cells — a breakthrough on the way to resurrecting the northern white rhino.

The zoological park at Höör in Skåne, southern Sweden, is home to a small herd of Heck cattle. These animals were bred in efforts to bring back the aurochs, but in the view of modern scientists the attempt fell short of its aim.

In southern Sweden, the aurochs could grow to a colossal size — up to 1.8 metres at the withers. This bull was discovered in Skåne and is now preserved in Lund University's zoological collections.

Nikita Zimov stands in the abundant grass of Pleistocene Park, surrounded by a swarm of mosquitoes. Grazing herbivores keep the land free of trees and bushes, which would otherwise cover everything.

Sergey has acquired an old Soviet armoured personnel carrier to knock down trees in the park — until such time as there are mammoths available to do the job.

The grazing herbivores and the open landscape are conducive to flowering plants. Beautiful pink carnations grow everywhere.

The foot of a deinonychus, the dinosaur on which *Jurassic Park*'s velociraptors were based. Although palaeontologists sometimes find amazingly well-preserved dinosaur fossils, there is no hope that they can be used to research the creatures' genetic makeup.

Models of the skulls of a velociraptor and an archaeopteryx, the first bird.
The velociraptor was about 50 centimetres tall, while the archaeopteryx was about
the size of a magpie.

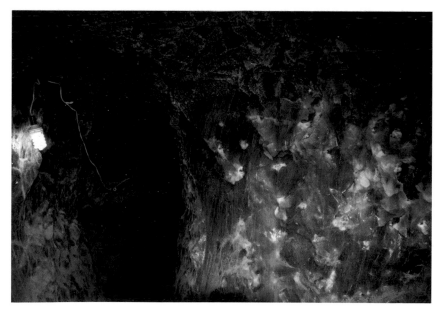

In Siberia, people dig caves in the permafrost to keep food and other things in cold
storage during the summer. A few metres down, the temperature is a constant minus
nine degrees, regardless of whether it is high summer or winter at the surface.

Arkhat Abzhanov has succeeded in modifying the development of chickens in the egg, causing them to develop a dinosaur's snout instead of a beak. However, he is doing this for the sole purpose of gaining insight into evolution and has no intention of recreating any dinosaurs.

At Duvanny Yar, the permafrost is thawing in a slow landslide that lasts throughout the summer. Mammoth bones emerge and trees lose their grip when the land under them turns into slurry.

The melting permafrost also reveals layers of pure ice trapped in the soil. When these melt, the earth above them collapses, and new lakes form.

The view of the landscape from the research station is very attractive. This is one of the most beautiful places I have seen. But Sergey is less impressed. People need to realise this is not nature, but a cemetery, he says.

Yet there's reason to hope the American chestnut will make a comeback.

'In another five years or so, we plan to start planting out blight-resistant chestnuts in the forests again. The only thing left is the legal approval process,' says William Powell, a scientist at the State University of New York who has spent the last 25 years trying to develop a blight-resistant chestnut tree. Cheerful and enthusiastic, William breaks into laughter from time to time during our conversation. He seems to be genuinely delighted that yet another journalist wants to interview him about his beloved chestnuts.

'When we started, we thought we'd cover this in five years, and of course it turned into 25. You know — young faculty and over-enthusiasm,' he jokes. His involvement with chestnut trees began when he was a doctoral student, though it took another few years for this project to get under way. Now he has only ten years to go to his pension. The trees he hopes to plant in the next five years are his life's work.

For quite some time — since the early 1980s — a project that involves crossing American chestnuts with a blight-resistant Asian species has been in progress. The aim is to produce a hybrid like the American chestnut but resistant to disease. But Asian chestnuts are much smaller than the American variety, so, ever since the first hybrid was created, scientists have been backcrossing various hybrids with American chestnuts, trying to minimise the share of genetic material inherited from Japanese trees, while preserving the resistance trait. This is a laborious task, as many hybrids inherit undesirable genes from the parent trees. Sometimes, for example, crossing produces a small, non-resistant tree — not a big, resistant one.

William's approach was different. He began by searching for a trait that would enable the chestnut to shield itself against the parasite. When the fungus first takes hold under the bark, it starts producing oxalic acid (the substance that makes wood sorrel and rhubarb taste sour) to kill off the tree's tissues. William identified a gene in wheat that enables the wheat plant to neutralise acid and thus to resist the fungus that produces it. Wheat has this gene because the plethora of fungal diseases that deploy the acid technique forces many plants to develop defensive mechanisms.

'This is a very common gene in plants. It's found in strawberries, bananas, and other things, as well as wheat.'

This is the gene he has spliced into the genome of the chestnut tree, using the same method that scientists apply to create genetically modified crops. Initially, William considered taking genes from the Japanese chestnut, thinking they might be more effective, as the two species are so closely related. However, the Japanese chestnut's resistance to blight depends on several genes, so it was easier to identify a single gene in a different species that would do the job. In addition to the wheat gene, the team have also spliced in a genetic marker enabling them to check whether the change has really taken effect and had the desired impact. William calls the latest variant of his resistant chestnut 'Darling 54'.

'We've managed to produce a type of chestnut that's even more resistant than the Asian variety. People ask me whether I've created a new species, but that's not the way it is. When you cross two different trees and get a hybrid, you've got a new species, but this is a much smaller change than you'd get by crossing.'

He shows me a film of an experiment in which three separate groups of plants are infected with fungus at a very early stage in their growth. The leaves of the ordinary American chestnuts roll up, shrivel, turn grey, and fall off. The Asian chestnuts do rather better, though the plants are smaller. Their leaves begin to droop a little and turn pale yellow. Some of the plants look as though they're dying, while others look almost healthy. In between stands the group marked 'Darling 54'. These plants have shot up and are covered in luxuriant dark-green foliage, a sign of health that could hardly be clearer. Darling 54 is the best variant so far, and William is clearly proud of the result.

He is just embarking on the legal process that has to be gone through before he and his colleagues can plant out the genetically modified trees in the open. This is the process that GM crops have to undergo in the United States before they can be planted in the fields. It's a laborious procedure and will take between three and five years, but, once it's over, the trees can be planted anywhere in the country. Today, the scientists are allowed to grow genetically modified trees for research purposes in a limited number of places. They have to clip the male catkins off these trees or tie bags over them, to stop any pollen escaping into the natural environment.

In the wild, it takes between seven and eight years for the trees to start producing chestnuts, but the team have managed to speed up the process in the laboratory and their plantations. There were not many from this first harvest, and they have all been sent to a laboratory to have their nutrient content analysed.

'We're being very, very careful with this — we want to make sure of success,' William says.

The hope is that the next nuts to be harvested can be planted and produce more seedlings. William is quite sure the team will get the approval they need. There's nothing to suggest otherwise.

William has tried to examine the role the modified trees

would play in a forest. He wants to rule out any unintended consequences. The team has looked into a wide range of issues, such as whether the insects that feed on the flowers are affected in any way, and how the leaves that fall to the ground are broken down into leaf mould.

More than likely, many creatures will benefit from the trees' flowers and nuts, but William is clear that there will be losers, too. He just doesn't know which species will be affected. In their experimental plantations, the team have found an unusual beetle that depends on the chestnut tree. Presumably, it will become more common if the trees spread, like other species surviving on the stumps that remain in today's forests. But more chestnuts will presumably mean fewer oaks, so there's a risk that oak-dependent species may decline.

Planting out transgenic trees and letting them spread in the open is a sensitive matter. There is strong resistance to genetically modified organisms, and it's common for activists to sabotage research sites. I ask William whether there has been any criticism of the project from environmentalists or other organisations.

'Not that much, really. I give a lot of talks each year to explain what we're doing, and I haven't come up against that much criticism. The trees we've planted on the campus haven't been damaged either. So I think people can see there are some good uses for genetic engineering.'

Yet it's hard to say how people will react when William and his fellow scientists start planting out trees in the forests.

The aim is for the transgenic chestnut trees to do precisely what worries so many people about GM crops. The scientists want the new genes to spread to wild trees. William hopes the trees they plant will crossbreed with as many as possible of the specimens that are still just about surviving in the forest. Most of the 'living dead' trees in the forest wither away before they can produce nuts or pollen, but now and then it does happen. If they were to crossbreed with the transgenic trees, blight-resistance would spread to their offspring. That would enable the species to recover.

'There's still some genetic diversity in the stumps that are left. We want to try and rescue them by getting them to crossbreed with the transgenic trees. Our aim is to make use of the diversity that still exists in the wild, because the transgenic trees are genetically all very similar.'

The gene the scientists have spliced into the chestnut trees is dominant, so a chestnut seedling has only to inherit that gene from its parents for the resultant tree to be blight-resistant.

The blight-resistance William has engineered doesn't mean the fungus dies. Rather, it can continue to live within the tree, just as with Asian chestnuts. Surely that's a problem, I say; won't it just make things worse if the fungus can spread at any time?

'Not at all,' William replies. 'Actually, it means there's less risk of competition between the fungus and the host trees. If a genetic change in the tree kills off the fungus, there's huge evolutionary pressure on the fungus to evolve and find a way around the problem. If the fungus is neutralised instead, but it can carry on living, there's less risk it will find a way to overcome the tree's resistance.

'You can add more genes to reduce the risk even further. We're thinking about a few different options. But in practice, there's probably very little risk the fungus will become resistant.'

All this means the American chestnut has stolen a march on all other projects to resurrect or recreate extinct species. If the application for a permit is successful, the new chestnuts will be the first revived species to spread again, though it can of course be argued that American chestnuts never really died out completely.

At the moment, William and his fellow scientists are waiting for the regulators to approve the next stage, subject to studies of how the trees affect insects and other organisms in their environment. But his team is already planning to start cultivating the seedlings to be planted out once the permit comes through. The idea is to run the project on a not-for-profit basis. There are no patents on the genetic changes they've introduced, and they plan to sell their plants at cost price. Their aim is to have 10,000 plants ready for dispatch as soon as they are given the green light. The idea is that anyone can plant chestnut trees, possibly in their own gardens, though the team really hopes the trees will be planted out in the forests again.

Abandoned open-cast mines are one option. Mining companies have a duty to restore the natural environment when mining is finished. William thinks such sites would be ideal places to plant both chestnuts and other species of trees. There are also plenty of abandoned fields and patches of waste land currently returning to nature that he thinks would be suitable for chestnut trees.

'People ask me: "So what are you going to do? Are you going to go out and cut down other trees so you can plant chestnuts?" The answer is: absolutely not! But there are plenty of places where forests are making a comeback, and there are always openings in forests where old trees have been brought down by a tornado or a small fire. Those are places where you can start planting chestnuts.'

There will probably need to be a great many chestnuts in the forests before they can begin to spread by natural means;

exactly how many is unclear. One difficulty with establishing the new population is that so many forest-dwellers like to feed on chestnuts. If there are only a few trees, nearly all the nuts will be eaten by squirrels and insects, rather than growing into new plants. Once there are enough trees, however, at least a few will manage to reproduce.

'But by then I'll be retiring, and I'll get a couple of acres of land and plant a few chestnut trees on it,' William laughs.

When will it all be finished? I ask. How long will it be until chestnuts can sustain themselves without human support? How long will it be till there are enough to be felled and used for timber once more?

'You know, I haven't actually thought about that, because I'll probably be dead by then,' says William, laughing again. 'I always tell people we're going to start this restoration project, but it's a century-long project. It's going to take a big effort. If we don't have people to plant these trees out, they're not going to spread on their own. You might have a few yard trees here and there,' he continues, rather more seriously.

He still sees the wound in the forests where the chestnuts disappeared a hundred years ago, the species that have struggled to survive. This is an ongoing crisis that he hopes he can use his trees to resolve. His approach offers a way to restore something infinitely precious that we have almost — but not quite — lost. What's at stake is not just the trees themselves, but all the effects they have on the ecosystem. In this sense, the project resembles Ben's dream of bringing back the passenger pigeon, though the image of majestic trees covered in white blooms is far less alarming than that of huge flocks of pigeons sweeping across the countryside, devouring vegetation and leaving dung behind.

I believe William will make his dream come true. For one thing, the chestnuts disappeared so recently that there are powerful scientific and biological arguments for restoring them to the forests. It looks as if the trees will do well, and in all probability they will benefit many other species, too. Above all, though, I believe this project will be a success because it is so easy to love trees. There are already committed volunteers who are keen to help plant them out, and, when it comes to goodwill, it's hard to beat a project to restore land devastated by open-cut mining. This goodwill is strong enough to convince even sceptics of genetic engineering. I admit, I am among those who have fallen in love with the idea of restoring the American chestnut to life.

Listening to William talking about the catastrophic loss of the United States' chestnut trees, I can't help thinking about the forests of Sweden. Death has visited Sweden, too — several times, in fact. Dutch elm disease, another fungal infection spread by beetles, has destroyed vast quantities of beautiful trees, in Sweden and the rest of Europe. While a healthy elm lives for four or five hundred years, an elm with this deadly disease can die in a matter of months.

'Nearly all of Sweden's elms are infected. But I don't think elms will disappear altogether, as they often manage to grow into saplings and disperse their seeds before they become infected. So I think the species will survive, but it'll change; there won't be any big trees any more,' says Johanna Witzell, a researcher at the Swedish University of Agricultural Sciences in Lund who specialises in fungi that grow on trees.

The fungus is so widespread that the island of Gotland is now the only part of Sweden where anyone is even trying to combat it.

Since 1997, the European Union has supported a project to try to preserve genetic variation in elms and assess whether there are any disease-resistant trees. The participants have collected hundreds of clones of trees from all over Europe in efforts to find a possible way of helping European elms recover.

'A lot of work has gone into producing disease-resistant hybrids and clones. Many of them are already available commercially. The question is: can these hybrids replace woodland elms, for instance, in terms of their ecological role? That's something I'm not at all sure about,' says Johanna. The hybrids are 'small and square', she says, more suited to parks and gardens.

Moreover, it looks as if the very fact that they're resistant to fungal diseases may be problematic. Trees are covered in multiple types of fungi that live within and on them, just as we humans harbour millions of bacteria. These fungi fulfil a function in that they affect various processes within the tree. Johanna's research has shown that elms that are resistant to Dutch elm disease also host fewer species of fungi in general.

'We need to ask what will happen to these disease-resistant trees in the natural environment. When they die and it's time for them to rot, this may involve different processes with fungi and bacteria. It could have a cascade effect on the ecosystem.'

Dutch elm disease is not the only condition to affect trees in Sweden. Ash trees in the south of the country are currently dying of a fungal infection first identified in Poland in 1992. Ash dieback (caused by *Hymenoscyphus fraxineus*) attacks shoots, making them wither and die. At the moment, there's no effective cure for the disease, and nothing can be done to protect infected trees. The disease has spread rapidly. In Sweden, it was first discovered on the island of Öland, but ever since 2005 the fungus has been found in all the areas where ashes grow. It looks very much as if it will kill most of Sweden's ash trees.

However, scientists in Denmark have discovered that a small proportion of ash trees are resistant to the disease. This means there's some hope that the species will survive and resistant trees can begin to spread, perhaps with human help. Although there are no firm plans so far, some European scientists think the disease-resistant trees that have been identified could be used to develop a new population. Scientists in France and Germany are urging that samples be collected from the remaining ash trees, both those that have survived infection and those that are uninfected.

William thinks another possible approach would be the one he's taken: identifying genes in other species that could provide protection against both ash dieback and Dutch elm disease.

'My project is definitely relevant to ash dieback and Dutch elm disease. But the question is: are people prepared to accept genetically engineered trees in Europe? I don't know if they are; they seem more worried about it than people here in the United States.'

By 2009, a quarter of all ash trees in Sweden were either dead already or seriously affected. And the disease has taken hold even more since then. The damp woodlands where ash trees grow are doubly threatened because virtually all the elms that once grew among them have already succumbed to Dutch elm disease. If the ash disappears, a whole range of other species will disappear along with it.

'I think genetic engineering is a great way to restore species without making a big change to them. But the caveat there is you probably only want to use it if you think you're going to lose all your trees. Essentially, you're starting from scratch, you're making a resistant tree and you're restoring, not preventing the disease,' says William, who doesn't see genetic manipulation as a panacea.

Johanna doesn't see it as a credible solution.

'I think it's far too slow and too dependent on chance. It's not as if genes are the solution to everything; there are other factors involved. I don't think this approach would be effective enough, and it costs too much in relation to the benefits it could bring. That's my view, and I've been working in plant biology since the 1990s, when people were just beginning to talk about the opportunities gene technology offered.'

The main problem, in her view, is that approaches such as gene editing take too long and arrive far too late in the day, once the diseases concerned are well-established.

'I'd go to the root of the problem. As far as Dutch elm disease is concerned, one of the reasons for the epidemic is the fact that we've used genetically identical elm clones everywhere in forestry. That creates ideal conditions for disease. So I'd begin by changing the way we use trees and forests. But then we'd need to scale back our expectations as regards the economic role of forests. I think we need to work towards increasing their underlying genetic diversity.'

But does that mean we should just accept the damage these diseases can cause?

'Maybe we need to accept the fact that forests are going to look different; there are other invasive diseases that are going to change our woods. We may be on the way to losing our great beeches, by the way, as they're succumbing to another disease.'

She starts telling me about the fungal disease that has just begun to threaten Sweden's beech forests, and that really worries me. Having grown up in the northern part of Skåne, Sweden's southernmost region, I feel as if I've spent half my life surrounded by beeches. I love these majestic forests, both in the gold of autumn and in the almost-other-worldly light-green of spring. Johanna and I are in her study at Alnarp, a campus of the University of Agricultural Sciences outside Lund, with a park full of colossal beech trees.

The disease now posing a threat is another fungus, which grows in the soil and is related to the potato blight that led to the Irish famine of the mid-19th century. The fungus spreads through the earth, attacking tree roots and hindering the absorption of water and nutrients. It's already present in the soil at Alnarp, Johanna tells me, and has started to weaken the trees, depleting their foliage.

'When we found out that the beeches in Malmö's Pildammsparken were showing this sort of damage, a lot of worried people rang us. Of course this is bad news. We've only looked into it on a limited scale, but we're finding the fungus everywhere; Söderåsen National Park is full of it, for instance. It's very hard to say what will happen to the forests with this sort of damage, as it's a slow process, and the trees may be more resistant than we think now. But if the worst comes to the worst, it could be very serious.

'But I hope there's enough genetic variation within our tree populations for there to be some tolerance, so not all the trees will die, as some of the people we talked to feared. I hope we can avoid the worst-case scenario,' Johanna continues.

Six years have passed since scientists first discovered the disease, but it has almost certainly existed here for longer.

'We're always a few steps behind when it comes to forest pathology and damage to woodland. Research doesn't get under way till the damage is already visible and it's started to have a big impact. By the time forests are that badly affected, it's generally too late. It's terribly difficult to eradicate these diseases; we can't get at them, and they're very hard to fight,' says Johanna.

The only solution is to stop such infections entering the country in the first place.

'As regards damage to forests and tree-based diseases, the international trade in plants is one of the main sources of new pests which our native species can't fight off. The trade in plants should just be stopped — there's simply nothing else to be done if you want to avoid these problems.

'No one's prepared to pay for a plant that's been screened, or for one that's been grown in a way that ensures it's free of disease — that pushes the costs up so much. But you really can do something as an individual; you can ask where plants come from, avoid buying from big plant nurseries in Germany or the Netherlands, and try to buy plants grown here in Sweden. That's a way to curb the spread of disease to some extent.

'If things carry on as they are, we'll have a lot more pests and invasive species from other parts of the world. So I hope people can develop a more long-term view and be more prepared to accept lower profits. It's probably very naive of me to say so, but I'd like to believe it could happen. If we carry on pushing the system to its limits, there's not much hope.'

I ask her what she thinks Sweden's forests will look like in the future.

'I think they'll be much younger; there are quite a few forces pushing forests in that direction. With diseases and other factors, we just won't have the same ancient woodland we have now. For instance, we won't be able to cultivate such old or such extensive beech or oak forests as those we've had up to now.'

CHAPTER 10

If It Walks Like a Duck and Quacks Like a Duck — Is It an Aurochs?

When the Red Army was marching on Berlin, Hermann Göring is said to have gone out to his country estate, Carinhall, and personally shot his cattle to prevent their falling into Russian hands. His sense of priorities may strike us as being more than usually unhinged for a man about to lose a war. Presumably, Göring was convinced that he was acting in the best interests of the Aryans — the Aryan breed of cattle, that is, rather than some human variety. The thing is, he believed his cattle were aurochs.

Nearly 15,000 years ago, at the end of the last Ice Age, when the thick glaciers retreated and melted, broadleaved forests gradually spread to cover large areas of Europe. These were not dense forests, but park-like areas with numerous open spaces and quite large areas of steppe. They were full of animals. In southern Sweden, for example, there were small mammoths, giant deer, musk oxen, wild horses, European bison, and the majestic aurochs.

Like many other Swedes, I first encountered an aurochs in the shape of Mura, the cute, shaggy cow from *The Hedenhös Children*, a series of books about a Stone Age family. In real life, aurochs were probably rather intimidating. They varied in size across Europe, achieving the greatest stature in southern

Scandinavia and northern Germany. Aurochs bulls could reach 1.8 metres at the shoulders and weigh as much as one-and-a-half tonnes, while the cows were slightly smaller. Their horns, which could be up to a metre in length, were light-coloured, with a darker tip. The aurochs had short hair, the bulls being blackish-brown, while the cows were more reddish in colour.

When the glaciers retreated, successive waves of people spread across Europe, all of them keen hunters. While the mammoths and giant deer disappeared rapidly, the aurochs and bison remained. Several thousand years later, people began to domesticate the aurochs in the regions that are now Turkey and Pakistan, and most probably in North Africa, too. The two or three lineages of tame aurochs that emerged were the forebears of all the cattle alive today. Once domesticated, they became smaller and more biddable, and they were bred to reach sexual maturity earlier, to calve from an earlier age, and to have more offspring. As a result of long-term selective breeding, today's cows grow rapidly and give more milk. But they are infinitely less resilient than their forebears in terms of defending themselves against wolves or surviving cold winters.

Gradually, as the European landscape evolved, with great forests giving way to towns and farmland, the remaining wild aurochs were displaced into remote areas. By the 13th century, they were confined to the eastern part of the continent: Poland, Moldavia, Transylvania, and Lithuania. They kept going longest in Poland, where, in the 16th century, the king decreed that farmers were to put out hay to help them survive in winter. Hunting aurochs was the prerogative of the nobility, and ultimately only royalty retained that right; poaching was punishable by death. By 1564, a head count in Poland put their numbers at just 38. Despite efforts to protect them, numbers dwindled fast. The last bull died around 1620, and one of his horns was turned into a hunting horn and presented to King Sigismund III of Poland. Today, it can be seen at the Royal Swedish Armoury in Stockholm, thanks to the propensity of the Swedish armed forces to help themselves to attractive artefacts while on campaign.

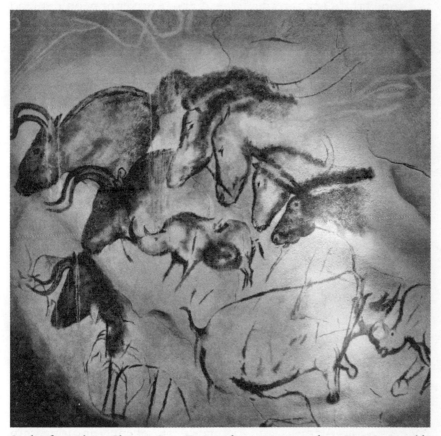

Study of aurochs in Chauvet Cave, France; the paintings are about 31,000 years old.

The very last aurochs, a cow, died in 1627. That made the aurochs the first species whose extinction was recorded. It would be less than 40 years until the next documented extinction — of the dodo. Although there were pictures of the aurochs, and references to it in various texts, the animal all but disappeared from memory as well. In the 18th century, scientists even debated whether the aurochs and the European bison, still living in the Polish forests at that time, were actually the same species. Or perhaps the aurochs had never really existed?

The European bison (also known as the wisent) nearly met the same fate as the aurochs. They managed to survive in the wild in the great forests of Poland until the First World War. Then Poland was occupied by Germany, and German soldiers based in forested areas shot over 600 animals. The last wild European bison was shot by poachers in 1927. At that time, there were some 50 individuals still living in zoos worldwide. Scientists set up a breeding program, enabling the species to be re-established from these 50 animals. European bison have been reintroduced into forests in Poland and a few other countries.

At about the same time as the last wild European bison disappeared, interest in the extinct aurochs began to grow. In the early 1920s, two German brothers, Heinz and Lutz Heck, began to dream of aurochs, inspired by old paintings and the mighty skeletons of bulls discovered in bogs in various parts of Europe. Both were directors of German zoos. Together, they decided to try to recreate the aurochs.

Their method was as simple as it was logical; since all modern cows and bulls were descended from the aurochs, the characteristics of that animal must be present in the hereditary material of cattle. They needed only to isolate the relevant parts. Heredity, and everything to do with it, was very much part of the zeitgeist in the Europe of the time; terms such as 'genes' and 'genetics' were still novelties, while the DNA molecule would not be discovered until 1953. Although interest in science was very strong in the early 20th century, there were huge gaps in understanding of how genetics worked by comparison with today.

The term 'landrace' refers to traditionally adapted and domesticated breeds. The Heck brothers decided that each would select a number of European landraces, the ones they considered most authentic and closest to the ideal aurochs, and crossbreed them. Their aim was to try to produce a purebred aurochs of the original type by eliminating the traits specific to the various breeds used. It was a kind of distillation process designed to wash away 10,000 years of breeding. They used a wide variety of animals, ranging from long-haired Scottish Highland cattle to Spanish fighting bulls. In 1934, both brothers announced that they had succeeded in breeding their own strains of aurochs. The experiments garnered praise, and the brothers proudly showed off their separate results. Various sources say their animals looked quite unlike each other, but that doesn't seem to have struck the Hecks as a major shortcoming.

By this time, Hitler had become chancellor, and — unsurprisingly — the Nazis were attracted by the idea of using selective breeding to reconstruct a proud, powerful breed of European origin. Lutz Heck was appointed head of the Third Reich's forestry authority, and the Nazi aurochs made an appearance in the regime's propaganda. Hermann Göring, who took a particularly strong interest in the project, had Heck cattle installed both on his hunting estate in today's Poland and at Carinhall, the country estate that lay north of Berlin.

When the fortunes of war turned against Germany, most of the brothers' 'aurochs' also died, either in conscious efforts to stop them falling into enemy hands, or in the general chaos. The breed created by Lutz Heck disappeared altogether, but some of Heinz

Heck's animals survived in zoos and nature reserves. Today, there are some 3,000 Heck cattle worldwide. They are allowed to roam free in nature reserves, where they keep the undergrowth down by grazing and maintain an open landscape. They are resilient animals, thriving on their own in the wild.

In the 1950s, people began to question whether the Hecks' cattle were really aurochs. Scientists concurred that what they had produced was not a revived species, but, quite simply, a new breed. The animals are too small, their coats are the wrong colour, and their horns are the wrong shape to resemble the original aurochs. Yet this doesn't mean that the dream of the aurochs is dead.

'My goal is to have herds of aurochs roaming free in large nature reserves in Europe,' says Henri Kerkdijk-Otten, the man behind the Uruz project. 'Uruz' is an ancient Germanic word for the aurochs. In practice, he has the same aim as the Heck brothers, but he hopes for better results, thanks to more modern knowledge of genetics — and without any Nazi delusions of European supremacy.

Henri, a historian who grew up in the Dutch countryside, has always had a fondness for cattle. Africa, and the wealth of wildlife there, are other abiding interests. He enthuses about all the animals you can see on safari, and the decisive impact they have on the landscape. He has a vision of a Europe as rich in big animals as the African savannah. Now he's trying to combine all these interests — one way being by trying to breed a new aurochs.

'For me, this is a way to recreate something we've lost and to restore the ecological processes associated with it. They were big animals, and they had a huge impact on the environment here in Europe, just as animals like elephants affect their environment, the savannah.'

One difference between this project and many of the others seeking to restore extinct animals is that Henri has less interest in genetic analyses. There's no need for the new aurochs to have exactly the same genetic make-up as their ancient forebears, he thinks.

'The thing is, you can analyse the genetic material all you like, but we still don't know which part of the genome controls which characteristics. It would be wonderful if we could say: "Let's take the gene for long horns from one breed of cattle and transfer the horns to the head of another breed." But there's no way of telling which genes or which section of the genome would actually do that.

'We want to create an aurochs for the 21st century,' he continues.

That sounds rather too much like a slogan for me to be able to take it without a pinch of salt. What he means is this: what matters is for the new animals to fulfil their function now, to look right, and to display the right traits. Whether they are carbon copies of extinct animals is less important. He tells me there are plans to compare the animals he and his associates breed with genetic analyses of aurochs bones, though this is not a crucial issue.

However, the animals' appearance does matter, as one of the ways in which Henri aims to finance the whole project is by marketing the meat of most of the bullocks as organic 'wild' meat. In this respect, their appearance is crucial; customers will want to feel they're buying genuine aurochs meat, says Henri. The meat will be sold while the breeding process is under way and he has the animals in relatively small enclosures, more like ordinary cattle. A lot of bullocks will be killed — along with various other animals not included in the breeding program — to ensure that only those with the desired characteristics can pass on their genes.

When I ask how the actual breeding process is going to work, Henri's enthusiasm is unstoppable; he starts to talk about a whole range of landrace breeds, their traits, and whether they have what it takes to help create a new aurochs. Apparently, Europe is teeming with landraces of cattle. He tells me they have selected four breeds that together have what's needed to make the new aurochs big enough, give it long enough horns and the right colouring, and ensure that its behaviour will enable it to survive in the wild.

But there's a competitive aspect to all this. Henri is by no means alone in dreaming of a new aurochs. Uruz is not even his first aurochs project. It turns out there are a plethora of projects that are all designed — using methods of varying complexity — to bring back the European aurochs. The only people intending to apply sophisticated gene technology, however, are Polish scientists who want to extract DNA from aurochs bones in museums and use it to recreate a specimen, just as George Church aims to recreate the mammoth.

A number of projects designed to take the Heck brothers' work further are also under way. The first was the German Taurus Project, which has been crossbreeding Heck cattle with other landrace breeds since 1996. Animals from that project can be found roaming around Denmark's Lille Vildmose nature reserve, and in a park in Hungary. A newer German project goes by the name Aurerrind and tries to combine several different breeds.

Then there's the Dutch Tauros Project, which, despite the similar name, has nothing to do with the German one. This project works in exactly the same way as the Heck brothers, crossbreeding various landraces in efforts to produce a new aurochs. The first crossbreeds are now being born, and there are breeding herds in various parts of Europe. In January 2018, a conference in Germany tried to bring the different breeding projects together for more cooperation.

The Tauros Project was where Henri began, but he chose to leave the project and set up his own, as he thought they were crossing too many different breeds. The last straw was when they started breeding with the long-haired Scottish Highland cattle, he says.

'They're small and fluffy, about as different from an aurochs as you can possibly imagine. They're the worst breed you could possibly have in a project like this,' he says sharply.

Henri clearly has very firm opinions on cattle. So he set up Uruz instead — a project that has gone for the most minimalist approach of all. By limiting the project to four breeds, he hopes to avoid wide-ranging, divergent genetic variation. The problem with involving too many breeds, he explains, is that though the resultant specimens may look right, they may still bear undesired genes. This means there will always be new calves with the wrong traits, and the breed will lack the necessary genetic homogeneity.

'With Heck cattle, you never know what a calf's going to look like; there are so many different genes washing around in their gene pool,' he says.

Henri thinks this year's matings may produce a calf with the right appearance — but that is not enough in itself. As he sees it, there are two aspects to breeding. Collecting together the right genes and mutations to give the animals all the features that characterised the aurochs is one, but the really hard part is sifting out all the other features he doesn't want. This is where he thinks all other attempts have failed.

The basis for his new aurochs will be the Italian Chianina, the world's largest cattle breed. They are impressive beasts, just as big as the aurochs of ancient times, but a beautiful milky white rather than dark brown. They also have quite short horns. Henri aims to get the shape of the horns from the African Watusi cattle. These have alarmingly long horns, like the Texas Longhorn. The other two breeds he wants to use are the Italian Maremmana and the Spanish Sayaguesa.

Once they've managed to produce an animal that looks right, the next step will be to release herds of the new aurochs in national parks in countries such as Spain and Romania, where they will essentially be left to their own devices. Some cattle will be shot from time to time, to stop them multiplying too fast and to prevent those that are not aurochs-like in appearance from passing on their genes.

It's one thing for these animals to look the part, I interject, but surely the hard thing is to make sure they behave in the right way, if they're to survive independently in the wild? According to Henri, that is actually less of a problem than one might think. It will happen automatically, as long as we release them into the wild.

'Take the herds of cows left on islands off the coast of Scotland. Ten years later, when people returned to the islands, the cattle wouldn't let them anywhere near. They attacked or fled, just like wild cattle — and that was after just a short space of time.

'We want to have an exact copy in terms of appearance, but we also want animals that function properly in ecological terms. So we work together a good deal with a national park in southern Spain, for instance, where we can release herds of aurochs and wild horses, and we're working with two national parks in Romania, as well as with northern Spain and a few places in Germany.' The idea is that the animals should be able to roam around semi-wild, and learn how to coexist with other large animals in environments that are relatively sheltered.

'With cattle, it's more about the environment than heredity,' Henri says confidently.

Whether he's right remains a moot point, but a fine estate in northern England may hold part of the answer. These days, Chillingham Castle is marketed as the country's most haunted castle; its most celebrated ghost is the shimmering spectre of a boy in blue. But the estate also offers something more interesting than ghostly apparitions — cattle.

Since early medieval times, a big herd of cattle has roamed free in a large enclosed area around the castle, and for at least the last three centuries they have mated only within the herd. Although extremely inbred these days, they seem to have avoided the genetic problems usually associated with inbreeding. The animals are completely wild and have developed various behavioural characteristics that set them apart from other cattle, which may support Henri's theory. For instance, they display more varied social behaviour than tame cattle, and they low to one another more. They have developed a hierarchical structure in which a single bull dominates the herd for a few years when he's at his strongest. During this time, he fathers virtually all the calves born. Young bulls are chased away, and they loiter on the periphery until one of them succeeds in challenging and finally replacing the dominant bull. But these cattle have no human or animal predators, and they have no need to defend themselves or their calves.

This is one of the main potential problems with aurochs, especially if they are to be allowed to roam free in various nature reserves, as Henri hopes. There will be people around them, and possibly predators. Semi-wild animals will need to develop a completely different mode of behaviour. What would happen if they became aggressive towards people? Or if they defended their calves against curious visitors, using their long horns? Or if they just ended up on the wrong road at the wrong time?

Being quite nervous around ordinary cattle, I can't really say I'm tempted by the idea of bumping into a 1.8-metre aurochs while out on a woodland walk. The aggressive nature of Heck cattle — attributable to Heinz Heck's use of Spanish fighting bulls in his breeding program — has been known to cause problems. Even if this particular project involves no such breeds, aggressiveness is a risk Henri is aware of. He says the animals they breed must be able to tolerate people.

'If they're to live in the Netherlands or Germany, for example, we're talking about quite small wildlife parks, where people would be walking around among them with their kids. If children were tossed by cattle, the project would be finished.'

That, in my view, is to put it mildly.

Breeding animals only to keep them in enclosures would be a waste, Henri thinks. After all, the aim is to restore Europe's original environment, to create a landscape like the steppes and forests teeming with animal life that followed the Ice Age. The aurochs is a necessary part of that, he says, as it would change its surroundings by grazing. He wants to achieve the same kind of effect that Ben aims to bring about with his passenger pigeons.

Henri also runs an organisation called the True Nature Foundation, which aims to 'rewild' the natural environment in Europe using methods such as releasing the future aurochs, along with wild horses and water buffaloes. He and his supporters hope that areas of Europe that have been abandoned as farming shrinks can be transformed into nature reserves.

The goal Henri talks about always involves releasing aurochs, allowing them to live wild in Europe again, and using them as a means to restore the environment to a wilder state.

'An aurochs is a tool you can use to return nature to something closer to its original condition.'

If it looks like a duck, walks like a duck, and quacks like a duck — is it an aurochs? Compared with the other researchers I've talked to, Henri seems both more pragmatic and more of a dreamer. He has a soft spot for cattle and is clearly nostalgic for a bygone natural environment. But when it comes to aurochs, he's also able to take an instrumental view; his goal is to recreate something that works, not a museum piece to be put on show in animal parks.

The aurochs isn't the only species that biologists are trying to recreate through breeding. In 2015, news arrived from South Africa that researchers had managed to resurrect the quagga, a cousin of the zebra with stripes on the front of its body, whose name reflects the sound it makes. The last quaggas died out in the 1880s, as a result of human activity.

Reinhold Rau began trying to resurrect the animal in 1987, by breeding ordinary zebras to produce more quagga-like animals. Now, some 30 years later, a small herd of quagga-like creatures can be found cantering about near Cape Town. Scientist Eric Harley, the project leader, has told CNN: 'If we can retrieve the animals or retrieve at least the appearance of the quagga, then we can say we've righted a wrong.' In this case, too, the idea is to release the new quaggas into the wild sooner or later.

This particular project is not one I find particularly impressive.

There's nothing innovative in breeding animals to have a particular appearance. The question is: how different is the aurochs from the quagga? It looks as if Henri will manage to breed a new aurochs that is closer to the original than the Heck brothers' creation. It also looks as if it will be released into at least a few nature reserves; the cooperation agreements and structures are already in place.

This has been a hard chapter to write. Starting with Hermann Göring and going on to talk about race, breeding, and origin is a tricky matter, even when you are writing about cows, not people. Ultimately, the question is how convincing the new aurochs will be. Fifty years from now, will we view them as we view Heck cattle today — as the result of bizarre experiments by eccentric researchers — or as a practical means of reviving a species?

CHAPTER 11

A Wilder Europe

They make an impressive sight, the big bulls and cows trotting around in the animal park near Höör in Skåne, southern Sweden, with their thick, curved horns, and their curly dark coats gleaming in the spring light. They are powerfully built, their bodies almost rectangular, with massive shoulders and high withers. These are the result of the experiment the Heck brothers carried out in the early 20th century. Though I'm rather nervous of cattle, I find these ones immensely appealing to look at, with their big eyes and the curly locks covering their foreheads. Yet the breed's origins are often held against them.

'I don't know what people think — that the cows themselves are Nazis, or what. Can't say they've started to moo "Heil Hitler" yet, as far as I've noticed,' observes Danish biologist Uffe Gjøl Sørensen drily.

Uffe used to work as a consultant at the Lille Vildmose nature reserve in northern Denmark, where a herd of 'wild cattle' have roamed within a large enclosure since 2003. Having been essentially left to their own devices, they've already started behaving more like wild animals, he tells me. Though no longer involved in the project, Uffe visited the reserve not long ago.

'I went inside and stood next to a tree. The herd started to move down towards the edge of the wood, towards me. One of the cows was ahead of the others, and, when they were about 50 metres away, she scented me. At that point, she suddenly came to a stop, then she just stood and stared at me. The whole herd came to a standstill with her. Not one came any closer. When I tried to approach them from the side, so I could take some photos from a different angle, the whole herd turned and went away again. They were wary, and they didn't want to risk my getting any closer. It was great to see they'd adopted a natural, instinctive way of behaving.'

This small herd belongs to the Taurus Project I mentioned in the previous chapter. Its participants are trying to continue where the Heck brothers left off, crossing Heck cattle with other landrace breeds to get closer to the original aurochs. Just two years after these animals had been released in the reserve, you could see the impact on the natural environment, according to Uffe.

'The area where they were released was almost completely overgrown, but, once the cattle were there, it became far more open. As I walked around, I could see it was beginning to form a mosaic of relatively wet areas and drier patches. The wetter areas were attracting lots of water fowl, lapwings, and so on. There are lots of beautiful butterflies there, too. It clearly suits them, having this kind of variety in the landscape all of a sudden.'

Uffe hopes that Heck cattle and other grazing herbivores can help re-establish the broadleaved forests that flourished in southern Scandinavia in Stone Age times. These were not the kind of deciduous forests a professional forester would approve of, with all the trees set very close together. Rather, they were extremely varied in structure. There were park-like open woodlands interspersed by wetlands and meadows; in some areas, the land was almost steppe-like, while the forests grew denser in others. The rivers and streams were constantly changing, fanning out into wetlands or being funnelled into narrower channels. Though scientists are still discussing exactly what the environment looked like at that time, they all seem to agree that there were broadleaved trees and large grazing herbivores in abundance.

In Uffe's view, this mosaic of a landscape is exactly what we should be trying to recreate. He would like to see trees in various stages of growth, open meadows, and bodies of water left to develop naturally, forming wetlands for a time or draining off through new channels.

'That variety is something we've lost. It's gone so completely that many people can't even imagine what it is we don't have any longer. We've got so far away from our original natural environment that it's very hard now to grasp what real nature is.'

According to Uffe, the big grazing herbivores like the aurochs and the European bison played a decisive role in shaping this landscape. That's why they're absolutely essential if we want to recreate it.

'The varied landscape was shaped by these large herbivores, and now — just like that — we don't have them anymore. If we try bringing them back, I'm sure we'll get a natural dynamic going that will be quite extraordinary. When I give my talks about the big herbivores, I use the title "Difficult to be with, impossible to do without". That's a line from a Danish poem which is actually about women — but as I say, I apply it to herbivores.'

'Difficult to be with' refers to the fact that these animals are challenging to work with, he says. They are large and heavy; the idea of driving into one is terrifying. They make demands on the natural environment and on the people living nearby. Given these major challenges, Uffe says they will clearly need to be kept in nature reserves.

'It's less about shutting the animals in, and more about giving them the freedom to do what they want, even if they have to live within a restricted area.'

I can only agree with him. Some voices in the debate are calling for animals like this to be released without any restrictions; they favour having a more genuinely wild population of animals in Europe again. I doubt whether Europe is quite ready for that yet. However, semi-feral grazing herbivores might be able to save the species-rich meadows that are currently becoming overgrown in various places.

The whole point is that these creatures change their surroundings, especially if they are released into forests full of dense underbrush. This change is what matters, even if it means living through periods when the situation apparently worsens, Uffe explains. He thinks existing efforts to protect nature in Denmark and Sweden have degenerated into something akin to stamp collecting, with people insisting that everything must look just the same year after year.

Imagine an area where rare orchids grow, he says. If you go there one summer and find 171 flowers, you'd like there to be 171 there next year, too. But then you release animals that graze on the plants in the meadows, including the orchids. The following year, there might be only ten orchids left, or maybe even none at all. That really doesn't look good. What people forget, though, is that orchids are adapted to surviving just such events. So if you come back a few years later, you'll find more orchids have appeared than there were to begin with. There will also be species of plants that weren't there previously.

'You have to take the long-term view. Over a number of years, you get a dynamic stability: a stability that works because of all the changes taking place in different directions all the time, a stability that creates a mosaic in a landscape which is constantly being reshaped.'

The story of the 171 orchids that have to bloom every year reflects a sense of insecurity among people working in nature conservation. They've seen so many species decline and disappear that their anxiety is hardly surprising, Uffe thinks.

'We can thank the people who loved counting orchids for all the natural environment we have left in Denmark today. The principle of fighting for what you hold dear, that's what's made the difference. Thanks to all that work, we still have something worth preserving today. But now we've reached a situation where we can try to take things further. Now we can start thinking about what sort of natural environment we really want.'

Restoring the European forests will take a long time, and grazing herbivores need to be used in an intelligent way, he reasons. Releasing them into a treeless area, for instance, would mean that no trees would have a chance of growing to maturity, as the herbivores would eat any new saplings. On the other hand, releasing them into an area full of mature trees would have very little impact. As he says, grazing animals can't climb trees to eat the leaves, so any impact they might have on a mature forest would be minimal.

'It would be really interesting to experiment with several different types of grass-eating herbivores in the same area, as each species affects the natural environment differently. That would pave the way for a much more varied type of environment. After a while, you'd suddenly see from the flowers, birds, and insects that something was really happening.'

The idea that you can recreate entire ecosystems by establishing a few species and allowing them to reshape the natural environment into something wilder and more authentic has grown tremendously in popularity over the last decade. The movement is known as 'rewilding', and there are now many projects worldwide that claim to be doing just that.

Grass-eating tortoises, for instance, have been released in a nature reserve in Hawaii as a substitute for the large grazing geese that were hunted to extinction. The aim is for the tortoises to eat invasive plants, just as the geese did, thereby creating an environment that is more favourable to native species. However, the outcome of this experiment remains unclear as yet.

In Sweden, beavers and wild boar provide examples of animals that had died out but have now returned and are having a major impact on the natural environment. Wild boar, which disappeared from Sweden in the 18th century, made a comeback in the 1980s when a number escaped from enclosures. They've multiplied rapidly, and in the forests of Småland, where I'm writing this, you can often see patches of soil where wild boar have been rooting around or broken up the ground with their trotters. This species has definitely started to make its mark on nature again, and research shows that their habit of rooting is beneficial to threatened plant species. On the other hand, wild boar cause thousands of traffic accidents every year, and the idea of these powerfully built creatures with their sharp tusks deters some people from walking in forests.

One of the aims of the Rewilding Europe organisation is to transform unused farmland into a wild environment with the help of horses, bison, and deer, particularly in southern and eastern Europe. Five regions have been selected to set up rewilding projects. The organisation draws attention to how a wild environment can attract tourists and provide a source of revenue for depopulated rural areas — in Spain, Croatia, and the Carpathian Mountains, for example. It has been criticised for its over-optimistic estimates of the amount of land in Europe that could be given over to such use. However, there is no need to exaggerate; the amount of land used for farming in Europe is quite clearly shrinking, and the forests are very much on the way back in countries such as France. According to Rewilding Europe, we thus have a perfect opportunity now to reshape the continent's wild environment.

In May 2014, the organisation released 17 European bison in the Carpathian Mountains in Romania. They are supposed to return to the wild completely, rather than being fed or managed by people. Rewilding Europe hopes their numbers will grow to 500 in the course of the next decade. The bison were kept in an enclosed area for a year, but in June 2015 the gates were opened, and the beasts are now completely free to roam. Rewilding Europe has also put a herd of animals from one of the many aurochs projects — Tauros — to graze alongside horses in an enclosed area of land in Croatia, hoping that they, too, will have an impact on nature.

Some of the variants of the rewilding idea seem very odd. Scientists in the United States have discussed the possibility of releasing into the wild elephants and camels imported from Africa, as substitutes for the megafauna that died out in North America about 10,000 years ago. They argue that since there are no close relatives of the animals that died out, the original landscape can only be recreated by importing exotic stand-ins. In Denmark, a few circus elephants were released into an enclosed nature park for a few days to see how they would affect the environment — supposedly standing in for the forest mammoths that once lived in the region.

The question of whether rewilding is a good idea, or even feasible, has led to a major debate among the world's biologists. One criticism is that what these projects are doing is creating a totally different new environment, not recreating an ancient one. There is, in fact, no real consensus about which era the various projects are designed to recreate. Another criticism touches on a more general problem: in the past, things have often gone very badly wrong as a result of humans trying to introduce animals into an area and predicting what would happen. It is precisely the wide-ranging effects that the 'rewilders' are trying to achieve — the impact of resurrected aurochs on nature, for instance — that their critics see as the main counter-argument. It's very hard to predict what such effects might be.

However, there is one example of a project designed to restore some of Europe's original natural environment that has been in place long enough for some conclusions to be drawn about its results.

Thousands of animals roam about on grassland crossed by streams. Herds of horses are grazing, with colts rearing up from time to time for a skirmish. Bulls and cows huddle in small groups with their calves, chewing the cud, while long-legged deer cleave to the trees. Waders make their way along the shore, foraging for insects, and a sea eagle appears, silhouetted against the sky.

The abundant animal life here is reminiscent of the African savannah, yet I'm in the Netherlands. This is Oostvaardersplassen, an area of wilderness reclaimed from the sea. Depending on who you ask, this is either a wonderful example of rewilding, or a failed experiment that has caused a huge amount of needless suffering.

But let's take a step back. This is Flevoland, a region of the Netherlands that lay under water for much of the last millennium. Land-reclamation work got under way with the building of a dike in the 1930s, and the emerging polder was dried out in the 1950s and 1960s. Now Flevoland is one of Europe's most fertile areas of farmland. At the end of the 1960s, an area of 56 square kilometres was set aside as a nature reserve, and a unique experiment began. It was an attempt to create an environment like that of the early European Stone Age — in a place with no terrestrial biological history whatsoever.

Heck cattle were brought in as a substitute for aurochs, while Konik horses were imported from Poland and red deer from Britain. The animals thrived and multiplied rapidly. Today, the area is a major tourist attraction. The wetlands attract birdwatchers, and you can drive out to view large herds of animals. Sea eagles arrived here of their own accord, the first pairs to brood in the Netherlands for several centuries. They've been held up as an example of how well the area has succeeded in creating a 'self-wilding' environment.

Initially, there were 20 horses and fewer than 60 deer. Now there over 1,000 of each. The Heck cattle have done less well; from the original 32, their numbers have grown to about 350. They are less successful than their rivals, the horses, presumably because cattle are less skilled at finding food in winter. And it is in winter that the major problems become visible.

The idea has always been to leave nature and the animals to their own devices, free of human intervention. However, the area has no predators, so the numbers of grazing herbivores have exploded. They fill the park and give birth to their young over the summer, but during the winter and early spring their food runs out, and the animals suffer from starvation. There are gruesome images from the park that show staggering, emaciated deer drowning, lacking the strength to cross shallow bodies of water; hundreds of corpses lying in the grass, waiting to be devoured by scavengers.

This situation has drawn protest from the general public, including calls for a number of animals to be culled each year, not in late winter when they are already on the verge of death by starvation, but in the autumn, to prevent the situation from arising in the first place.

Frans Vera, a biologist who was one of the park's founders and who has been very influential in its management, is strongly opposed to people going into the park to shoot animals. He has said in several interviews that he sees cycles of growth and starvation as natural processes; he thinks people should avoid applying their personal morality to animals. Moreover, he thinks that predators — the wolves now spreading across Europe from Germany and Poland — are bound to reach the area sooner or later. A wolf has been spotted nearby, though there are none in the park at the moment.

There are now over 5,000 animals in the park, and, according to certain estimates, as many as 2,000 could die in the next hard winter. Over the last few years, the authorities have decided to start shooting animals that are believed to lack the capacity to survive. Some critics say the cull often comes so late that large numbers of animals are left to suffer unnecessarily. Another option under discussion is the possibility of giving animals contraceptive injections to cut the birth rate and thus reduce the population.

As I write, a debate is raging in the Netherlands about how to manage the Oostvaardersplassen in future and who should bear ultimate responsibility for the animals and their wellbeing. The many articles on the subject in Dutch newspapers make it clear that this is a toxic issue. The political parties have very different ideas about the area's future. At the time of writing, it's not clear exactly what will happen, but it looks as if there may be drastic changes in the way the park is run.

'We humans must go in and play the role of predators; we just have to. This project isn't sustainable in the long term,' says Uffe. In his view, the animals in the park need to be culled. And those who carry out the cull must select their quarry in the same way as an animal predator would, which implies killing either old or very young animals — although that would cause an upset among the park's visitors. No one wants to see playful calves or frolicking young deer being shot, yet that's closer to the way a genuine predator would behave, says Uffe.

Everyone I talk to says the best solution, and the most difficult one, would be to bring in a vigorous group of predators to hunt the big herbivores.

The advantage of animal predators over humans is that they change the environment in ways other than simply keeping numbers down. The wolves living in the American national park of Yellowstone have apparently even changed the course of the rivers. The area's original wolves were hunted to extinction in the early 20th century, but about 20 years ago biologists released a small group of Canadian wolves into the park.

Yellowstone's wapiti (also known as elk) were directly affected by the wolves. Their numbers fell, and, above all, they became more wary; it seems they now prefer to keep out in the open. This has meant that certain areas are more closely cropped, while others are less affected. As a result, shoots and saplings have been able to grow rapidly into bushes and trees without being eaten immediately; the roots of these bushes and trees have affected the soil, and one of the rivers running through the park has changed course owing to the changes in vegetation. Overall, the landscape has diversified, enabling more small birds and rodents to thrive. In their turn, predatory birds, foxes, and other animals have benefited.

The wolves of Yellowstone are often held up as a perfect example of the importance of predators, and of what can happen in nature if all the pieces of the puzzle are in place. That interpretation, on the other hand, has been criticised as an over-simplistic description of the various processes that interact in a natural area, and people have questioned how many of the changes are really attributable to the wolves.

Anyone who has followed the debate about wolves in Sweden will know that the presence of predators is not without its problems, and there are similar debates going on in all the other places where predators have been reintroduced by scientists or have returned of their own accord. This is certainly true of Yellowstone, despite the positive impact that wolves have had on the natural environment. It isn't easy to get people to accept predators in their vicinity, not even if they are in enclosures.

Predators are the main problem in all the rewilding projects currently under way. Not only is it trickier to gain public acceptance, it's also harder to raise and release predators without their becoming too used to humans and adapting their behaviour.

Despite this, there are now more predators in Europe than there have been for a good many years. Germany has a large wolf population. A wolf wandered into Denmark almost exactly 300 years to the day after the last Danish wolf was shot; today, there are about 40 in the country, and the latest reports say at least two pairs are breeding. The Danish animals made their way across from Germany; Uffe teases me, arguing that the present Danish wolf population is genetically more resilient and more diverse than the Swedish one.

Regardless of whether Heck cattle, European bison, or resurrected aurochs are released in Europe, the continent looks set to become wilder in the years to come. For several decades now, European forests have been gaining ground at the same time as farmed land has dwindled. These are not just forests managed for forestry purposes, but what might be called new wild forests, left to their own devices.

A wilder Europe is almost certainly on the way, but it will not be without its problems.

CHAPTER 12

'Most People Would Call This Totally Insane'

Nikita Zimov guides the boat more cautiously as the water becomes shallower. We've been travelling by motorboat for several hours through the maze of broad, shallow stretches of water around the research station in Chersky, Siberia. Suddenly the boat stops, and Nikita leaps out and starts towing it. I imagine we've run into a sandbank and he's pulling us clear, which has already happened a few times. But he carries on wading, towing the boat in his wake. After a while, he tells me to get out, too.

Nikita gives me the once-over, appraising the boots I'm wearing and my short stature, and says, 'I'd better take the camera and your rucksack now.' This turns out to be a smart move. I ease myself gradually over the gunwales, boots first. 'Don't jump,' Nikita says. I'm expecting to feel some kind of sand or gravel underfoot, but there's only mud that yields beneath my boots, giving me little purchase. The river water runs straight in over the tops of my boots. Nikita, who has trainers on, is oblivious of his wet feet. He supports me as we wade towards the shore through what feels like bottomless sludge under a shallow layer of river water.

From a distance, the shore looks enticing, covered in light-green wavy grass like a summery meadow inviting you to go for a gentle run. In reality, it's composed of tussocks nearly a metre high, crowned by bright-green grass sprouting on top of a layer of dead grass and grass roots. Each tussock is about 30 centimetres across. It's impossible to walk over these tussocks, which are unstable and wobbly, and thicker at the top, like chicken legs. They are spaced just far enough apart to ensure that they provide no collective support; instead, they bend towards each other as soon as you attempt to put your weight on one. In between them lies mud, sometimes just ten centimetres deep, though generally my whole boot sinks in so far that dark-brown ooze seeps all the way down to my toes. The clumps of bright-green grass are like an unruly lock of hair, making it hard to see where I'm going. I keep treading on the edges of the tufts, which bend and slip underfoot. I must look like a drunkard as I make my way across, constantly putting my feet in the wrong place and wobbling from side to side.

My boots are full of water, and I fall over again and again, putting my hands on the wobbly tussocks to steady myself. Nikita is striding forth, far ahead of me; he isn't even making any visible effort. I, on the other hand, am sweating and panting and swaying as I stumble onwards. I can't even be bothered to try to bat off the aggressive mosquitoes.

Gradually, the land rises among the tufts, and the mud thickens from a syrupy consistency to something more like thick porridge. Walking becomes marginally easier. Some of the tussocks have been grazed here, and we spot a dark silhouette in the distance, clearly the one and only bison. Thanks to him, the musk ox, and the small herd of horses, the land here is beginning to resemble a meadow, making it possible for me to stand up and look around rather than keeping my gaze constantly fixed on the ground.

All the shores we passed on our way here were covered in virtually impenetrable thickets of sallow and other bushes. In the slightly drier areas, the bushes are replaced by dense larch forests. This is the only patch of open grassland I've seen that isn't a marsh. It's a result of the great experiment that Sergey and Nikita are carrying out in this remote area. They call it 'Pleistocene Park' — a nod to another park with natural phenomena brought back to life.

The Pleistocene was a geological epoch that began about two-and-a-half million years ago, giving way just over 11,000 years ago to the Holocene, the age we live in now. Essentially, the park's name refers to the end of the Pleistocene. At that time, Siberia was a thriving steppe landscape, home to mammoths, woolly rhinoceroses, bison, asses, horses, wolves, and more. It was dry, windy, and dusty, but the grass was sufficiently abundant to support almost as many animals as the African savannah.

When people arrived, both the open landscape and much of the fauna disappeared. It is still unclear what exactly happened. The climate changed and the Ice Age came to an end, but the steppe landscape here had already coped with major climate change in the past without losing all its mammoths and other fauna. Scientists are engaged in a vigorous debate over whether early humans are the reason why Siberia is now covered in species-poor forests, not steppes teeming with life. Nikita and Sergey are convinced that this is the explanation.

Their hypothesis is that there were enough hunters to tip the ecosystem out of kilter. Grassland needs herbivores to graze it, or it becomes overgrown with underbrush and scrub. So when numbers of herbivores fell as a result of hunting, this had a knock-on effect, with the land becoming overgrown. That meant there was less grass available for the remaining animals. This vicious circle was perpetuated by humans, who may also have found it easier to spot their quarry in the few remaining areas of steppe. The result was the forests and thickets that cover today's landscape. At all events, that is Sergey's interpretation of what happened. Today, the forests are home to a few elk, but not many other animals apart from the big, plump ground squirrels that are to be seen everywhere.

Yesterday evening, I asked the elder Zimov why the big animals never returned. Siberia is sparsely populated, and even without mammoths and woolly rhinoceroses, large populations of grazing reindeer and horses should have been able to transform the landscape into grassy steppes again. Sergey says that even though there are so few people in Siberia, there have always been enough hunters to prevent the fauna from recovering. Periods of recovery have been followed by spells with more intensive hunting. One example, he says, is what happened to the reindeer after the fall of the Soviet Union. There was no longer enough funding available for commercial reindeer herding, so the sector went into a decline. Instead, wild reindeer began to appear, animals that would once been hunted by reindeer herders, but which now spread rapidly, even reaching Chersky. According to Sergey, there had been no wild reindeer here for a century, but suddenly there were several thousand of them.

'But a few people got their hands on some cash and invested in two high-quality snow scooters so they could go hunting in winter. Within one year, about 30 people managed to kill all the wild reindeer — maybe as many as 15,000.'

This has recurred time and again throughout history, he says: a few hunters have managed to wipe out, or at least to decimate, whole populations of wild animals. It has been that way ever since humans first arrived here, says Sergey; that's why the steppe has never recovered.

'There's always been sallow and grass along the riverbanks, enough grazing for as many as several million animals. But there've also been a few thousand people — enough to prevent those animals from ever multiplying.'

Out in the meadow, Nikita and I stand watching the small herd of horses grazing not far away. The aim of Pleistocene Park is to try to recreate the Siberian steppe landscape with the help of grazing herbivores. The enclosure contains about 40 horses, a few musk oxen, a lone bison, a small herd of reindeer, and a few elk. They have been grazing here for almost 20 years, and the difference in the landscape within and outside the enclosure is striking, with abundant grass and open areas inside, and thickets outside. The enclosed land may possibly be beginning to resemble the landscape that once existed. The park opened officially in 1996, and Nikita took over the reins in 2004, at the age of 20.

'The thing is, my dad can't be doing with all the paperwork, so he's handed it all over to me. In other words, I'm not really the one in charge — but if anyone were to land in jail, I'd be the one,' he laughs.

One of the Zimovs' aims is to bring more animals here and expand the area of enclosed land, to make the animals' impact more visible. The problem is transporting the animals to this area, particularly musk oxen and bison, neither of which live anywhere nearby. Nikita tells breathtaking stories about driving lorry-loads of animals over the frozen tundra for weeks on end, during the short period in spring when the weather is not cold enough to kill them, but the ground is still frozen, and the bogs and rivers covered in ice. There are no roads to this area; the only way to get here is by driving across country.

'I forced myself to drive for 17 or 18 hours a day. I'd sleep for something like four hours, and spend the rest of the time taking care of the animals and eating. The lorry was brand new when I bought it at the start of the journey, but everything was broken by the time I arrived. I had no brakes and no lights left. I drove over ice as it was cracking. I really had to step on it to reach ice that was still bearing. It was incredibly frightening and exhausting.'

He goes on to tell me how the shaggy musk ox got here. Nikita and Sergey took their boat out into the northern Arctic Ocean without any decent navigation equipment. After a week or so, they managed to reach Wrangel Island. This island, once home to the very last mammoths, is now a nature reserve full of musk oxen. Nikita and Sergey had been promised that they could take a herd of young animals back to the park, and the individuals set aside for them had been placed in an enclosure. But when they arrived, it turned out that a polar bear had broken in and killed some of the musk oxen, while the rest had run off.

'So we had to spend ten days trying to find the herds, anaesthetising the young animals, and transporting them to the boat. When we'd managed to catch six of them, they all turned out to be males. But by that time, we had to go home.'

It was a stormy homeward voyage, and Nikita stood at the helm for two nights in a row.

'I had a chart, and there was GPS, but I had absolutely no idea what we had in front of us, and I had to navigate among icebergs in the pitch dark. You really don't want to run into an iceberg and re-enact the *Titanic*.'

Only one of the six male musk oxen is still alive, and, without any females, there won't be any more. Maybe they should set out on a new expedition soon to collect more animals, but Nikita is rather hesitant, given all the difficulties and dangers it would involve.

There are enough animals in the park to keep the bushes down and allow the grass to grow, but they can't do anything about the trees in the forest. Nikita and I hoist ourselves into a small six-wheeled all-terrain vehicle and start driving along something that one might — with a considerable stretch of the imagination — call a forest track. I have to hold on tight with both hands to stay on board as the vehicle rumbles over the bumpy ground. Finally, we drive off the track, straight into a group of larches with trunks ten centimetres in diameter. The trees snap like matchsticks around the vehicle, and Nikita switches the engine off.

'Most people would call what I've just done totally insane.'

There is a pungent scent of resin from the broken trunks. I'm dizzy after the drive through the forest and the abrupt end. The silence is overwhelming after the roaring of the engine, and the mosquitoes find us within a few seconds. It occurs to me that 'most people' might conceivably have a point.

Nikita begins to talk about how the forest is an intruder here; this area should hardly have any trees at all. He talks about how modern people's view of nature is misconceived. People think the forest is nature at its best. Yet it is actually the least favourable form of nature, he says. It isn't natural at all, merely what happens when grazing herbivores are eliminated. And this is where the mammoths come in; they once played a decisive role in maintaining the steppe landscape. The mammoth was the only animal in the ecosystem large enough to break and kill trees, thus enabling grassland to spread. Nikita and Sergey can release horses, bison, musk oxen, and other animals here, but at the moment there is nothing to replace the mammoth.

'This is our mammoth calf,' Nikita jokes, patting the bonnet of the six-wheeled vehicle. Sergey has also got hold of a superannuated Soviet armoured personnel carrier, which, in the absence of any adult mammoths, he drives around the park to knock over trees. Both men think that knocking over trees is the biggest contribution a resurrected mammoth could make.

So I ask Sergey whether he thinks it's necessary for George Church, or someone else, to recreate mammoths so they can re-establish a steppe ecosystem in Siberia. It happens to be Sergey's 60th birthday today, and he responds with an apt analogy for the occasion.

'You know, this morning my wife asked whether she should bake me a cake. I said it'd be nice, but not absolutely necessary. You can always celebrate your birthday without having a cake. It's the same with mammoths. Of course, it would be good to have them again, but this is going to work, either with or without them. It'll take longer for the forests to disappear, but it'll happen in the end.'

Sergey is a little sceptical about the various projects working to recreate animals through genetic engineering. He regards them more as PR stunts than as serious research. The only project he's really keen on is the plan to bring back the European aurochs. He thinks it very likely to succeed and sees it as essential for Europe's natural environment. Nikita is more positive about the possibility of recreating mammoths.

'Yes, there's definitely a need for mammoths. The problem is just that it's going to take a while. George is a very smart person, and he's doing a good job. They're probably going to be able to produce a mammoth lookalike some time very soon. But it's not one mammoth we need — what we want is a herd. Establishing a large group of mammoths will take a long time. Creating a herd that's big enough for them to start developing social and behavioural patterns is a process that may well take a hundred years.'

George's ultimate aim is to release a group of mammoths in Pleistocene Park, as he told me when I met him in Boston.

'My long-term goal — though it's not something I'm obsessed by like Captain Ahab by the white whale — is for there to be 100,000 cold-resistant elephants in Siberia, Canada, and Alaska. It would be the same process as when we went from nearly zero bison in the United States to the half-million we have today.'

He hopes that if that happens the mammoths — together with horses, reindeer, and other animals — will change the landscape here, recreating the Siberian steppe. That is just what the Zimovs are trying to achieve.

'I always look for a project that's cool in philosophical terms, one that'll drive technological development, and that has benefits for society,' George says. He's convinced that modified Asian elephants in Siberia would benefit all those involved, from the natural environment to the local people, who would be able to earn money from mammoth tourism.

Even the ever-present mosquitoes would be less troublesome if the forest were replaced by grass, Nikita thinks. Rain and snowfall are quite low up here, but as water is prevented from draining away by the permafrost, it accumulates in shallow pools and sluggish rivers, which make ideal breeding grounds for mosquitoes. The theory is that fast-growing grass absorbs more water than slow-growing trees, so the ground here would dry out if the forest were replaced by grassland.

Nikita hopes the ecosystem they create will have the same wealth of animal life that flourished during the Ice Age, and that they can develop the same complex interaction between groups of different species.

'After 10,000 years of farming, we humans still haven't learned how to manage grazing land as effectively as animals within ecosystems have done on their own. So what we're trying to show is how we might be able to look after our planet in a way that's far more effective and productive than today — and show how productive such ecosystems can be even without fossil fuels. Productive in terms of food for us humans, too.'

It would be quite possible to hunt in a system like this, he thinks, provided that hunting was restricted to a low level. The major problem at the moment, according to Nikita, is how to introduce predators when the herbivores grow in numbers. There are bears, wolverines, and wolves here, but, of those three, only wolves can attack big animals. Predators or hunters are necessary to prevent the herbivores from over-grazing, ruining the ground, and dying of starvation in spring, as has happened in Oostvaardersplassen. How exactly they are going to solve that problem remains unclear at the moment.

'The commonest misapprehension about us and our research is that we're crazy,' says Nikita. 'But I don't think I'm crazy in the least. I'm very pragmatic: I look after the park, I'm trying to save the ecosystem here, but I'm not doing it out of pure idealism. I want to create a good product that will generate profit — maybe not money, but other forms of profit that will benefit people and humanity in centuries to come.'

CHAPTER 13

A Chicken's Inner Dinosaur

I can't imagine a single reader has got this far without wondering, in some frustration, why — apart from a brief reference in the introduction — I haven't yet mentioned either dinosaurs or *Jurassic Park*. This film, which came out in 1993, is guaranteed to be the first thing to come to mind in connection with resurrecting extinct species. It has come to stand for both the potential and the risks of de-extinction.

So iconic has *Jurassic Park* become that everyone thinks they know just how to go about bringing dinosaurs back to life. In the movie, the scientists take a fine piece of amber containing a perfectly preserved mosquito, drill a tiny hole, and extract the blood that formed the bloated mosquito's last meal. They manage to extract the dinosaur's genetic material from the dregs of blood, and sequence its genome — and then all they have to do is set about producing a dinosaur egg.

Real-life researchers have, in fact, tried to do that very thing, searching for genes in insects that are astonishingly well preserved in amber. This involves crushing the amber in a laboratory kept perfectly sterile to avoid contamination by any other vestiges of genetic material — the scientists themselves wear spacesuit-like garments — and then attempting to extract minute pieces of

DNA molecules from the dead insect. This is the method Beth Shapiro is using to sequence the mammoth gene from frozen bone fragments. She has also attempted to extract DNA from insects in amber.

The problem is this: scientists can't find any dinosaur DNA in mosquitoes. They can't even find any DNA from the mosquitoes themselves, or from any of the millions of bacteria that must have lived on them once. In short, they can't find any DNA at all. One test appeared to show the remnants of some insect genes. However, when it was repeated, the researchers discovered that the sample had been contaminated by something in the laboratory; it had picked up a dab of genetic material from a fly or some other insect.

Given the DNA molecule's fragility, the oldest creature from which scientists have extracted fragments of sequenceable genetic material is 700,000 years old. That is quite remarkable, considering how new the technique is. Still, it remains a far cry from the 65 million years that a dinosaur genome would have had to survive to be analysed.

Scientists examining particularly well-preserved dinosaur fossils have managed to identify remnants of a handful of proteins, including collagen, keratin, and elastin, but they have not found even the tiniest fragment of DNA. Some scientists have even gone as far as to suggest that DNA might have a kind of half-life, meaning that it would break down even under optimum conditions. This would make it theoretically impossible to find out what the genes of a tyrannosaurus looked like and how they differed from those of a stegosaurus.

However, that doesn't mean that the dream of being able to see a dinosaur one day is dead. There really are scientists who are working to achieve just that, though not in the way you might imagine.

There are times when a journalist stumbles across characters who seem too good to be true. Jack Horner is one such. He found his first dinosaur bones at the age of eight and decided to become a palaeontologist. He dug up his first sizeable dinosaur aged 13 and has continued to make spectacular finds ever since. Despite having serious dyslexia, he has developed a successful scientific career. His discoveries have played a decisive role in giving scientists the understanding of dinosaurs they have today. For instance, Jack was the first to discover that these creatures built nests to lay their eggs in, nurtured their young, and lived in herds — a far cry from earlier views of dinosaurs as clumsy, primitive dimwits. On top of that, he has had at least two dinosaurs named after him.

Jack was also the model for Dr Grant in the *Jurassic Park* films — the cowboy-hatted hero who saves the children and all the other characters when the giant saurians escape from their enclosure and start eating everything and anything they can lay their claws on. Much of the dialogue in the various films refers to Jack's own scientific discoveries.

Jack also acted as scientific adviser on all four *Jurassic Park* movies, working with Steven Spielberg, the cast, and the team that developed the special effects to make the dinosaurs as lifelike as possible. His link with Hollywood goes even further: producer George Lucas is now funding his dinosaur resurrection project. As I say, it all sounds too good to be true.

'The main reason I'm doing this is because I can. There really is a chance we may be able to build a dinosaur,' he says.

Our conversation is punctuated by chuckling and chortling on Jack's part. He clearly thrives as much on hypothetical issues and philosophical discussion as he does on media attention. He seems tickled when I ask critical questions about his dinosaur-restoration project and the difficulties he faces. In many ways, he resembles George Church, the man who aims to recreate the mammoth; both seem to be driven by the same kind of enthusiasm and curiosity, and they are of a similar age.

Since it's impossible to investigate the genetic make-up of the dinosaurs, which disappeared over 65 million years ago, Jack has had to find another way to create his miniature 'terrible lizard'. His plan is to take a chicken as the starting-point and to try to coax forth its inner saurian.

Biologically speaking, birds are dinosaurs. Not only are they descended from them, but they actually constitute a group within the dinosaurs' family tree, just as lions belong to the feline family and rats to the family of rodents. It's just that all the other branches of the tree have withered away. So from a purely scientific point of view, Jack could simply place a hen on a pedestal and say, 'Tada!' However, as he says, that argument wouldn't convince a sixth-grader keen to see a *Tyrannosaurus rex*. Birds may be a subgroup of dinosaurs, but it's not sparrows that come to mind when we hear the word.

To cut a long story short, there are four things that distinguish birds from other types of dinosaurs. Chickens have wings rather than arms and hands, and a beak instead of a snout. They are toothless and have a short, compact rump instead of the long tail characteristic of a dinosaur. Everything else, from their feathers to the wishbone people squabble over after a roast-chicken dinner, are features they share with the dinosaurs, Jack tells me. Dinosaurs belonged to a very diverse group of animals that lived over a very long period of time, whereas the group Jack compares modern birds with are therapods, the group from which birds evolved.

Therapods included species such as tyrannosaurs, velociraptors, and other two-legged dinosaurs with long, narrow heads and long tails.

'What we're trying to do is reverse the evolutionary process and get an embryonic chick to develop into a dinosaur instead.'

It may be possible to transform a chicken into a miniature dinosaur in the same way that George Church is trying to transform an elephant into a mammoth: by choosing to remove certain genes and simply replacing them by others. In this way, a hen could acquire teeth, just as an elephant calf could grow fur like a mammoth.

But Jack has opted for a quite different approach. For one thing, there is no mould for dinosaurs' hereditary material that scientists could use as a basis, and for another Jack is at least as interested in the processes as in the final outcome.

About 150 million years ago, the dinosaurs that would eventually evolve into today's birds began to diverge from the other saurians. Feathers were already common; many dinosaurs had down or ostrich-like plumes. But the group that would eventually evolve into birds developed wings and began to fly. The other changes that set birds apart from other dinosaurs came within a fairly short span of time. The first 'real' birds appeared about 100 million years later. What exactly happened during that time, and what process did birds' evolution follow? That is the question Jack wants to answer by recreating a dinosaur.

His team plans to take chicken embryos as a starting point and steer the development of the foetus, controlling the genes that are active while the embryonic chick is developing inside the egg. In this way, they aim to reverse 150 million years of evolution and produce a more archaic creature. The process resembles Henri

Kerkdijk-Otten's approach to back-breeding aurochs, though it is more complex by far.

The underlying idea is that, since birds had teeth, a tail, and other features at earlier stages of their evolution, the genes for these features still exist in their hereditary material. Evolution can be characterised as a process that constantly adds new features to existing ones, like layers in an archaeological dig. Genes for traits that have fallen into disuse may nonetheless remain present in hereditary material, like junk in an attic. Some of these features emerge at the embryonic stage, only to disappear again. For instance, all human embryos have a clearly defined tail for a number of weeks, but it disappears in time. These are the discarded genes that Jack aims to reactivate.

This involves no genetic engineering in the conventional sense. What the scientists need to do is steer embryonic development in a different direction. They seek to switch off the processes that result in features peculiar to birds and to replace them by older processes from the genetic archives. Jack thinks there's a chance they may be able to reverse and eliminate a series of evolutionary chances and actually make a creature that looks like a dinosaur.

Numerous scientists worldwide are trying to steer birds' embryonic development in efforts to understand evolution and see what dinosaur-like features they can bring out. Some focus on the beak, others on how arms and hands developed into wings. A few teams of scientists have managed to get chickens to develop rudimentary teeth. Jack, however, is the only person who wants to connect all these aspects — and who has openly declared that he aims to hatch a baby therapod.

He started planning the project several years ago and wrote a book called *How to Build a Dinosaur*, which came out in 2009. However, it took some time to arrange funding, and the laboratory has only been working actively on the project for about four years when I talk to him. The first stage has been to investigate how birds' rumps develop at the embryonic stage.

'At the moment, we're focusing on changing a bird's rump into a long tail, which is the hardest bit. It turned out no one knew much about birds' tails, so we've had to go a bit further back to find out what goes on inside the bird. We can't start reversing the process till we know how it actually works. That's rather slowed down our progress towards making a dinosaur, but we've found out lots of interesting stuff along the way,' Jack enthuses.

Once they have worked out how a bird's rump develops, they have to find a way of getting it to grow into a long tail at the embryonic stage. That would be a huge scientific breakthrough; it would entail getting the embryo to develop new vertebrae. Not only is such knowledge essential in engineering a dinosaur, but it could also be used to help people with spinal conditions. Yet Jack makes it clear that the project isn't designed to achieve such applications, which would just be a fortuitous by-product.

'The point of all this is to discover new stuff. I think our society is far too obsessed with the idea that all research has to have a purpose and be applicable to people's lives. I don't agree with that — I think the point of science is to find out as much as possible about the world, no matter whether or not we can benefit from that.'

Even if they succeeded with the tail, they would still be a long way from their goal. After that, they would have to collect all the research that has been done worldwide into the ways in which the other body parts evolved, and put everything together. Some of the research available would need to be taken further, too. The teeth which scientists have managed to develop in chickens, for example, are not particularly impressive. It might be possible to research certain aspects by looking at living birds. For instance, as a nestling, the South American hoatzin has something resembling fingers and claws; its embryonic development may be able to provide clues about dinosaurs' arms. However, certain aspects of embryonic development may well affect others, so there's no guarantee that the various components would fit together in a single chicken.

I ask Jack what it would take for him to feel he had completed his project.

'It would be when we can take a chicken — or any kind of bird, really — and activate genes so they produce teeth, change its mouth, give it a long tail, and alter its arms and hands. So in practice that would mean hatching a creature with a head like a dinosaur, with teeth in its mouth, arms and hands, and a long tail. It would look just like a miniature modern-day therapod.'

If he managed this, the animal would still be feathered, just like many dinosaurs, and it would be no bigger than a chicken. It would hardly be terrifying enough to land the leading role in a Hollywood movie, but its very existence would represent a huge scientific advance in terms of scientists' ability to steer embryonic development. How long does he think it will take before the first one hatches?

'It's nearly impossible to put a time-frame on it, because there's no way of telling how long each of the individual experiments will take. If we're extremely lucky, we can probably manage it within five years. If we're unlucky, it might take a decade longer. But even then, we're only talking about 10, 15 years, which isn't that long at all,' he says, looking cheerful and confident. No doubt about it, Jack is a born optimist.

Rows of hens' eggs lie in a Harvard laboratory, all with a tiny hole in the shell that lets you peek inside. They belong to another scientist who's trying to solve the same problem as Jack, but with a different aim.

'By examining how the foetus evolves, you can understand a number of things about evolution, and vice versa. I want to try to understand a very important change, the one that gave birds beaks,' says Arkhat Abzhanov, showing me into his study. The room is full of quantities of white skulls from different types of birds and a few models of dinosaur fossils.

He has succeeded in getting chick embryos to develop something that resembles a crocodile's snout instead of a beak. The scientific study on this was published in early 2015. Rather than modifying any genes, what he and the other researchers involved did was insert into the foetus a chemical that shut down certain components of its development. They changed the signals the growing foetus received. He shows me X-rays of the bones in the upper jaw of ordinary chick embryos, a crocodile, and the chickens he modified. Some of the modified embryos look far more like crocodiles than like ordinary chickens.

If you run your tongue over your palate, just behind your front teeth, you can feel the bone that Arkhat modified. You can feel the other bone if you pull your tongue further back in your mouth. At the highest point of the palate, you will feel an edge. In a bird, these two bones form the beak. These are the bones whose development Arkhat managed to stop and send into reverse in chick embryos, producing something closer to an embryonic dinosaur.

'This is quite a nice way to show what signals are needed for the foetus to form a beak. If we shut down the signals, the embryo falls back on an older program of foetal development. In practice, it starts trying to form a snout instead,' Arkhat says.

But he has no plans to have chickens with snouts wandering around his study.

'We've allowed the embryos to grow to quite advanced stages in their development, but we've never let them hatch. There would definitely have been ethical problems with that. In our project, we never let anything go as far as hatching.' The purpose of the project is to try to understand evolution, not to create dinosaur-like creatures.

'It's a very long time since modern birds developed. We have no idea how much of the genetic program in foetal development still remains and would work if we activated it.'

It's absolutely true that archaic genes and processes can remain in birds' genetic material, and that investigating them can enable researchers to discover important things about evolution. But that doesn't mean they are so well preserved that they would actually work in a live chick, he explains. A chicken that hatched with a modified beak, for instance, would have difficulty eating, since other parts of the skull are not steered by the same process and would not really be compatible with the new features. The embryos produced here lack a functional snout, he points out. The differences visible in the X-rays are just a way of enabling scientists to understand what took place during the evolutionary process.

You could describe this research as examining birds' family tree by moving backwards along the branches, he explains. We may be able to go all the way back to the point when birds diverged from the other therapods and understand what happened then. However, that doesn't mean we can reverse the whole process or that the genetic material concerned would be in good enough condition to develop a miniature dinosaur. DNA sequences that are not used tend to accumulate multiple harmful mutations; sometimes, large sequences disappear altogether.

Arkhat is extremely sceptical about Jack Horner's project.

'I think it's too early for anyone to claim they can bring back anything at all. All we can do is answer a number of very specific questions about evolution. I think even scientists are quite naive about how complex this is. The more we dig down into it, the more complicated it is.'

Arkhat guesses the best thing scientists will be able to achieve is an embryo that looks like a dinosaur, but can't develop to the point of being able to hatch and would be quite unable to grow to adulthood. In fact, he doesn't even want to engage in such speculation. It's too early and we know too little, he says. You have to bear in mind that even if a creature like a small dinosaur were ultimately created, there's no guarantee that its genes would function in the same way as in the original creature. Scientists might just have found another way of switching genes on and off that would produce the same end result.

The two scientists, working on chickens from opposite directions, have reached quite different conclusions. Jack is relaxed about Arkhat's critical and sceptical attitude.

'He's a brilliant scientist, but he's essentially a pessimist and I'm basically an optimist, that's probably the big difference between the two of us. To be honest, it may not work — who knows? It definitely won't work if we don't do these experiments. I'm ready to go through the whole process to put it to the test.'

Again and again in the course of the interview, he makes it clear that he doesn't know whether the project will be successful. He doesn't know whether it will be possible to hatch a dino-chicken, or whether this hypothetical creature would be able to grow to adulthood. He has no idea of what exactly it would look like, and so on. That's the whole point as far as he's concerned. These are things that he and his fellow scientists will find out as they go along. They are also things that they can't find out without actually carrying out the relevant experiments.

'Once we've figured out all the components and managed to combine them, we'll have an embryo. Then we have to get the chick to hatch somehow. Whether it'll survive or not, whether it'll be able to grow to adulthood, and so on — those are all things we don't know, but we'll have to find out by trial and error,' says Jack.

Setting aside all the huge scientific problems, Jack's work has also been criticised for being unethical. The creatures he is engineering would, in practice, be monsters, and they would suffer, say his critics.

'Of course we won't deliberately make an animal that would suffer. Whenever anyone brings up that issue, the first thing I do is remind them that we've bred dogs whose lower jaws are set so far back that their teeth stick out of their mouths. We've literally bred animals that suffer *by definition* compared with their ancestors, haven't we? Yet we keep them as pets. So can we say a bulldog suffers from being a bulldog?'

What sort of future does he see for his mini-dinosaur, if he manages to make one?

'The same future I see for a bulldog,' he says, and laughs. 'We've made all kinds of weird dogs and other creatures. Dinosaurs will be one of our domestic animals, and we'll be able to keep them as pets. If we wanted to, I think we'd be able to raise them for meat as well. I'm sure someone'll make a mint from selling them as pets, but it's not going to be me.'

There's no doubt that chicken-sized dinosaur pets would sell like hot cakes. They would represent a scientific breakthrough, but at the same time they would be delightfully harmless. There would be no risk of a Jurassic Park scenario, with dinosaurs turning on their creators and devouring them. Maybe they could hunt mice, just like pet cats.

Jack's boundless optimism is infectious. It's clear he's going to enjoy himself tremendously, no matter whether he succeeds or not. I hope he will, though I have enormous doubts. But Jack's reason for trying — to see whether it can be done — strikes me as being the best and most honest motive I've come across among all the people I've interviewed. Maybe it will work, maybe it won't. It's worth a try.

CHAPTER 14

The Fine Line Between Utopia and Dystopia

I began this book with the myth of Prometheus, who brought fire to humankind in defiance of the gods' interdiction. There are two readings of this story: you may think Prometheus acted well, or that his deed was mistaken. Tellingly, Prometheus is evoked time and again by those who view de-extinction as immoral and unethical.

'Attempting to revive lost species is in many ways a refusal to accept our moral and technological limits in nature ... things did not end very well for Prometheus,' writes Ben Minteer, professor of environmental ethics at the University of Arizona, and one of the more vocal critics of de-extinction.

In his view, we have a moral lesson to learn from witnessing species loss. The depletion of biodiversity reminds us of our own fallibility and limitations, protecting us from becoming intoxicated by the notion of our own power. He quotes American biologist Aldo Leopold, writing in the late 1930s:

'Our tools are better than we are, and grow better faster than we do. They suffice to crack the atom, to command the tides, but they do not suffice for the oldest task in human history, to live on a piece of land without spoiling it.'

The great challenge, in Ben's eyes, is not to revive extinct species, but to live more sustainably and start tackling the moral and cultural forces driving today's environmental degradation.

'We are a wickedly smart species, and occasionally a heroic and even exceptional one. But we are a species that often becomes mesmerised by its own power. It would be silly to deny the reality of that power. But we should also cherish and protect the capacity of nature, including those parts of nature that are no longer with us, to teach us something profound about the value of collective self-restraint and human limits. Few things teach us this sort of earthly modesty any more. It cuts against the progressive aims of science to say it, but there can be wisdom in taking our foot off the gas, in resisting the impulse to further control and manipulate, to fix nature.'

Ben's opinions are diametrically opposed to those of Stewart Brand. Stewart's goal is just such a future, in which we humans assume responsibility for and stewardship of nature. He would like us to play a more interventionist role, even in the regions we today regard as wild and pristine. While Ben wants us to put the brakes on and be mindful of our own fallibility, Stewart wants us to step on the accelerator before it's too late. Each sees his own approach as the way for humankind to limit the ongoing destruction of nature.

Stewart thinks we are already behaving like gods in the myriad ways we impinge on nature, and that it is better to be responsible gods than gods who destroy things by accident. Ben thinks it idiotic that we should even fancy ourselves to be godlike. Between them, they sum up the ethical and moral dilemma underlying any attempt to resurrect extinct creatures. This is no longer about whether gene technology works, or whether the new species could adversely affect the natural environment. It's about the inherent human dilemma.

I still find myself wavering between these two extremes, like Buridan's ass hesitating between two heaps of hay — though this is more like being trapped between two sabre-toothed tigers. I find Stewart's vision intimidating. While the natural world he envisages may be abundant and luxuriant, it resembles nothing as much as science fiction in its smooth, clinical rationality. And Ben's line of argument, with its fatalism and acceptance that the battle is already lost, is at least equally disturbing. His basic principle seems to be that we humans can't make the world any better, so we had best do nothing at all.

One specific aspect of what Ben says is particularly hard for me to accept: the concept of reducing the existence of other species to a moral lesson for humankind. I can't really accept the idea that we should give up trying to make the world better — both for ourselves and for other species — merely because the loss of nature is supposedly a lesson in humility. At the same time, I agree with him about our collective over-confidence in our own powers. It would be hard to disagree.

Should we refrain from taking action that would make the world better, on the grounds of our propensity to overestimate our own powers? Or, conversely, does species resurrection reflect human arrogance, rather than making the world a better place?

The problem is that ethical debates constantly culminate in dystopias or utopias, in all or nothing. That makes it harder to discuss the issues, and nearly impossible to reach a balanced overview.

One key aspect of this issue goes beyond what's happening in laboratories all over the world, or the question of technical feasibility. And that is what we humans, who would need to cohabit with any resurrected species, feel about them. It is, after all, our feelings that will determine what ultimately happens.

Susan Clayton is a professor of psychology whose specialist field is the study of how human beings relate to nature. When I talk to her, she stresses the fact that people value species diversity very highly. Most of them think we should do everything in our power to save threatened flora and fauna. At the same time, we value the wild and pristine quality of the wilderness, the fact that it's untouched by human hands. Releasing resurrected creatures into the wild could lead to a conflict between these two aspects.

'De-extinction makes me nervous because it's not just about individual species. It's something that changes the relationship between humans and the natural world,' says Susan. She's still undecided about whether species revival is a good or a bad thing. To shed further light on the problem, she begins describing the research conducted over the last 10 to 20 years into the psychological effects of spending time in natural settings.

'It has positive effects on people's mental health, their social experiences, and their cognitive skills. There's considerable evidence that it has a positive impact on wellbeing. But we're still trying to pin down why that is.'

There are various theories about this. One is that experiencing nature reduces stress; another is that our ability to direct our attention is restored in nature, which provides an environment that holds our attention without requiring us to make any particular effort.

Another possible reason for the positive effects we experience could be that we view nature as being free from human impact, as a setting in which human influence is limited. Susan tells me that many of the people she has spoken to experience a sense of humility, of being part of something bigger than themselves when they are out in the forest; they feel that human beings are not the centre of the universe, and this almost spiritual experience is very beneficial.

'I recognise that we humans do already intervene in nature in many ways: we have agriculture, we have genetic manipulation. But I feel that each new thing that increases human control over the natural environment might tend to reduce that sense that nature is a domain in which we can feel humbled and not so important. I have no evidence for that, so it's speculation.'

Resurrected animals could become a key symbol of how that relationship is changing.

'The concept of de-extinction implies a relationship in which humans have much more control over nature. It feels like a very big thing to change the relationship to such a degree that human beings get to decide which species we're going to actively recreate. That's very different from saving a species from extinction.'

As knowledge of the effect nature has on people is so new, it's impossible to say how our relationship with nature has changed in the course of history, or what direction things will take in the future.

'I think it'll change, and it probably already has changed. How much, I don't know; again, I think this goes back to not knowing why nature has the effects that it does. How much of it depends on actual reality, and how much on human perceptions? In 50 years, our experience of the natural environment could be very different, and we may have a strong sense of how we've changed nature — maybe through climate change, or in some other way,' Susan speculates. Such change could mean that we would benefit less from being outdoors in natural surroundings, though the reverse might be true.

Susan singles out another important aspect: 'There are trends in culture and ideologies, as well as trends in fashion. Right now, we're in an era of valuing everything perceived as being authentic and natural. But 50 years ago, things that were not natural, that were completely engineered by humans, were much more interesting to people. I'm sure the pendulum will swing back again at some point.'

So what view does she take of Prometheus and the perils of human hubris?

'I don't know. Being over-confident can be a good thing if it motivates you to try more, but it can also lead to big mistakes. I think being over-confident means you don't necessarily think through the possible drawbacks.'

At the same time, she thinks it vital that we assume greater responsibility for nature. Merely holding back and refraining from taking action could also have a very negative impact.

'Finding a balance there is the really difficult question if you try to say you want to protect nature, but not intervene in nature. But when it comes down to defining what *is* natural, the line is very, very fuzzy, perhaps even non-existent. So I don't have a good answer to that, apart from saying it's a very complex question, one that you need to be informed by biological knowledge, but also by psychological knowledge of how people might react.'

Many critics of the concept of de-extinction are anxious that it could make us care less about the animal species under threat today. This is a concern Susan shares.

'I'm worried it could affect the way people think about what's possible. They could start thinking, "We don't have to worry too much about protecting species, as we can always figure out some way to bring them back later."'

I ask her whether she thinks de-extinction could have the opposite effect, giving people hope and inspiring them with the idea of restoring animals that have died out. Might it not make people keener to save the species we still have today?

'Actually, that's a very good point. It's depressing to hear about species dying out the whole time, and many people report a feeling of burn-out. If you're very worried about nature and you feel that all the news is bad, it can make you give up. If you think you can't make a difference, why bother? But believing humans can tackle these problems could make you very motivated to do more, potentially, and could lead to more support for other conservation initiatives. And the fact that that could help inspire hope is one of the reasons I'm not taking a firm position.'

There's another important aspect that will affect how people respond to resurrected species. Will they really be viewed as being the same as the original species — or as entirely new ones? Biologically speaking, a recreated species is a completely new kind of organism, a variant of either an existing species, or of a lost one. But calling a hairy elephant a mammoth is a way of increasing acceptance.

'I think people would probably react differently and more negatively if scientists said they were creating new species of animals, whereas bringing back old ones sounds more benign,' says Susan.

Reviving something that once lived somehow seems both more reassuring and more morally acceptable than setting off on a voyage of discovery of the opportunities presented by gene technology.

'I think it's very easy for people to think that other people's behaviour is guided by rational information processing, but it's important to recognise the role of emotions in determining people's responses. People's behaviour is hugely affected by their emotional response, in many cases more than by their rational thinking about the issue.'

CHAPTER 15

A Melting Giant

'Can you smell the mammoth piss?' asks Nikita, delving into the black slurry formed by the melted permafrost. The whole shore is pervaded by an odour combining a whiff of the stable with a hint of pigs' manure, a smell whose pungency increases wherever the mud underfoot is so treacherous that you have to put out both hands to support yourself, to avoid falling over and landing with your face in the muck.

Starting at the Chersky research station, we have chugged up the broad, shallow River Kolyma for about three hours. This is Duvanny Yar, a site where the river cuts through the frozen earth, exposing a 40-metre-high bank of permafrost. Now, in July, the surface is thawing, and the melting ice embedded in the soil is transforming the whole bank into a slow cascade of mud. The site is a magnet for local mammoth hunters in search of valuable tusks to sell, but above all it attracts scientists from all over the world who come to study the permafrost, and what may happen once it starts to thaw.

Here in Siberia, just as in large areas of Canada and Alaska, the earth is frozen all year round, except for the topmost layer, nearly a metre in depth, which thaws for a few months each year. This is the layer where trees and grass extend their roots, where lemmings and ground squirrels (Siberian chipmunks) dig their burrows, and where all other biological activity takes place. Virtually nothing happens below that layer, and nothing has happened for thousands of years.

A thick layer of the special type of soil found here, *yedoma*, was built up during the last Ice Age. At a time when Scandinavia was covered by glaciers, this region was an open, grassy steppe populated by mammoths, woolly rhinoceroses, bison, and sabre-toothed tigers. The permafrost constantly yields up the bones of these animals; during our visit, we find a well-preserved mammoth tooth and a few bones that, Nikita guesses, come from a reindeer.

By collecting bones from along the shore, Nikita and Sergey have tried to calculate the numbers of animals that lived here when the steppes were at their most bountiful. They estimate that each square kilometre was home to one mammoth, five or six bison, six horses, and about 15 reindeer. It's hard to make estimates for animals smaller than reindeer or horses, because large bones are better preserved than small ones. There were also predators in the region, but they were fewer in number than the herbivores, and scientists haven't found very large quantities of their bones.

There are some problems with this method of estimating numbers, and the Zimovs may possibly have overestimated them, but it's clear from the bones found that this was once a rich environment full of grazing animals.

At that time, the region was dry, windy, and dusty. Year by year, dust from all over the world was blown here, where it settled in millimetre-thick layers on the grass. A new layer was added every year for 40,000 years. Because it was so cold, the soil froze from within and the ground rose gradually, while the ice swallowed up whatever it could reach from the underside. A glacier formed beneath the feet of the animals that lived here, and we are still walking around on top of it.

The soil here is rich in ammonia. One theory is that it comes from the urine of large animals which froze before it could decompose. Whole bodies are embedded in the soil, along with plant matter and other organic material. Nikita draws my attention to the thin threads in the soil that are the remnants of grass roots tens of thousands of years old.

'If you took all the vegetation on earth — forests, grassland, bush, everything — and put the carbon it contains in one scale, and then you put all the carbon in the earth's permafrost in the other scale, the permafrost scale would be more than twice as heavy,' he says. According to the most recent calculations, the permafrost contains about 1.3 trillion tonnes of carbon — one-and-a-half times the amount in the atmosphere. Sergey was one of the first scientists to show, in a groundbreaking article published in 2006, what enormous reserves of carbon are stored here.

This sequestered carbon is beginning to make the world's scientists and decision-makers increasingly nervous, because — just like everywhere else in the world — the Arctic is experiencing rising temperatures. In fact, it is actually warming up faster than the rest of the planet. Over the last 30 years, the temperature has risen by half a degree Celsius every decade. The main reason for this is that the decline in snow and ice has made the Arctic darker and therefore more liable to be warmed up by the sun's rays. Rising temperatures will also shift the tree line further north, so that areas that are light-coloured tundra today will be filled by darker trees and bushes, further amplifying the change. Conversely, the increase in vegetation will bring more evaporation, which will cool the land down to some extent, particularly in summer.

Despite the latter effect, the overall result will almost certainly be an increase in the amount of thawing permafrost. The carbon that is now sequestered in the ground will be digested by micro-organisms and released into the atmosphere as carbon dioxide or methane. According to the latest estimates, there's a risk that between 5 and 15 per cent of the carbon in the soil may thaw and be converted into greenhouse gases by the end of this century, if other greenhouse-gas emissions persist at their current rate. That corresponds to about a tenth of the total yearly anthropogenic emissions of greenhouse gases.

This is the real reason why the Zimovs want to re-establish the steppe landscape and George Church wants to release mammoths that graze and knock over trees. The project is an attempt to save the thawing permafrost, one that would prevent large quantities of carbon dioxide from being released into the atmosphere.

'How can we stop the permafrost melting? How can we keep this carbon frozen so it isn't released into the atmosphere? That's a huge challenge. But what we've shown with Pleistocene Park is that a steppe landscape full of grazing animals could be a simple, inexpensive solution,' Nikita says.

The reasoning behind this astounding claim is as follows. If dark scrub and trees can be replaced by light grass, more of the sun's warmth will be reflected back out into space during the summer. But the really big impact of a steppe full of grazing herbivores would come during the dark winters. Each winter, this part of Siberia is covered by a layer of snow between half a metre and a metre thick, which remains unmoved throughout the season.

'It's light, airy snow which provides good insulation. Even if the air temperature is minus 50 degrees Celsius, the snow prevents the cold from reaching the frozen ground. The temperature under the snow may be just minus ten to minus five degrees,' says Nikita.

That prevents the ground from cooling down as much as it might in wintertime. As a result, it doesn't take much for it to begin thawing once spring comes. So a solution could be to get rid of the snow, enabling the ground to cool down more over the winter and the cold to persist for more of the summer season. Less warmth would reach the permafrost, so it would stay frozen.

'Of course, you could send out thousands of snowploughs to shift the snow, so the cold would reach the permafrost properly. But setting aside the difficulty and the expense, the fuel would release so much greenhouse gas that it wouldn't make any difference.'

What he and Sergey have noticed is that grazing animals have exactly the same effect as snowploughs. They turn over the snow and compact it in their search for frozen grass to eat. This prevents it from insulating the soil. Scientists have tried to measure the impact by placing thermometers in the ground both within and outside the park.

'We took some measurements at the end of March, when it's coldest here. Half a metre down outside the enclosure, it was minus seven degrees. At the same depth inside the park, it was minus 24. That's 17 degrees' difference — just because of the animals grazing inside,' Nikita enthuses.

The hope of saving the permafrost inspired by this difference in the temperatures inside and outside the enclosure is enough for George Church. He says the possibility of keeping the Siberian permafrost frozen is one of the main reasons why he wants to resurrect the mammoth, rather than any other long-dead creature.

'That's a major reason for doing this on a large scale. Once we've produced the first animals, they'll live on whatever grows there. And once the process is under way, I think they'll be able to look after themselves. So it could be mammoths that save the climate.'

Further down in the permafrost, where the temperature is more or less constant, the difference between the soil inside and outside the enclosure is less marked, but Nikita is nonetheless convinced that this could be an effective way to protect the permafrost. It isn't only what would happen to the global climate if the permafrost melted that worries him. When the land starts to shift, it causes major problems locally. The school in Chersky had to close after a huge crack split the building down the middle. Trees become increasingly unstable and can fall over when the firm ground under them starts to subside. Roads and gas and oil pipelines are also at risk of destruction. In some places, the soil has subsided by nearly ten metres in just a few years.

A few days earlier, we were at a site that visiting American scientists have dubbed 'the hellhole' because there are more mosquitoes there than at any other spot. The forest there burned down about ten years ago; the remains of burnt trunks can still be seen everywhere. The fire was so intense that it burned off the top layer of insulating moss. As a result, the permafrost has recently started to thaw rapidly.

Permafrost often contains large quantities of frozen water, but it's not evenly distributed; instead, it's concentrated in lumps of almost pure ice. The ice here is in wedges that form a zigzag pattern around pillars of earth. When the permafrost begins to melt, the water drains off, while the earth remains where it is. The water can either drain out into rivers, or gather in lakes and other stretches of water, which in their turn accelerate the thawing process.

This has transformed 'the hellhole' into a virtually impassable tract of land. Apparently, not even one square kilometre has remained flat. It's composed entirely of steep hills and narrow oval lakes. The lakes grow deeper and the terrain becomes even harder to negotiate with every year that passes. Scientists say the land here was completely flat before the fire; today's landscape is entirely the result of the thawing permafrost. Variations on this process can be seen wherever the permafrost contains a large proportion of ice. The result is wounds in the earth that gape ever wider rather than healing.

The newly formed lakes cause a further problem. The oxygen-poor mud on the bed of a new lake is conducive to bacteria that produce methane, a greenhouse gas far more potent than carbon dioxide. Just how much of the carbon released by the thawing process risks being converted into methane is one of the biggest unanswered questions surrounding the issue of thawing permafrost.

The thawing permafrost is a problem so great as to defy comprehension, but a solution based on releasing millions of horses and a hundred thousand mammoths into the wild seems equally hard to grasp. There's a very big difference between showing that something works in an area just a few hectares in size and hoping to be able to put it into practice throughout the Arctic Circle. I ask Nikita if he really believes it will happen.

'I doubt whether the Russian, Canadian, Chinese, and American authorities will suddenly decide this is a good idea and start releasing animals. For the next 25 years or so, they're more likely to do very little. But sooner or later, there'll be a collapse, a point when a huge amount of permafrost will start thawing at the same time. Then the forest will die, because the trees will tip over when their roots lose their grip on the soil, as it turns into mud,' Nikita says.

'Within a short space of time, a lot more grass will grow — it's the first plant that appears on new land,' he continues. 'So within another 30 years or so, there'll be incredible amounts of grass for animals to graze on. By then, people may understand just how bad things are, and start releasing horses into the wild. The horses will multiply, and there'll be unlimited amounts of food for them, as all the ecosystems we know today will have collapsed. Within just a few years, if there are enough animals, they'll stop the permafrost thawing in the areas where they live, even if no one repairs the damage that's already been caused. Obviously, it would be better to get started right now, but ...' Nikita lets the sentence hang in the air, shrugging his shoulders in a way that expresses quite clearly what he thinks of slow-moving, unengaged politicians.

Of all the scientists I've spoken to, no one doubts that this approach would work. Grazing herbivores could almost certainly bring the temperature down enough to keep the permafrost frozen, even if the average global temperature rises. Even without mammoths to knock over trees, horses, bison, and other large herbivores would almost certainly have enough impact to make the vital difference.

The big — indeed, gigantic — problem with this solution to the problem of thawing permafrost is, of course, one of scale. Siberia is vast. Words are inadequate to convey its vastness. For nearly the whole duration of my flights from Moscow to Chersky — which took over ten hours in total — I was flying over uninhabited wilderness and the frozen wastes of Russia. A fifth of the land mass in the Northern Hemisphere is occupied by permafrost. Most of that area would need to be filled with grass and grazing herbivores to protect the frozen land.

There is a difference between theory and practice. There is a difference between enclosing an area just a few hectares in size and examining the impact there, and conducting the same experiment on the scale of a continent. Yet a restored steppe landscape full of grazing mammoths may be the only way to save the permafrost. The alternative would be to slow down global warming. However, even if the ambitious goals that came out of the 2015 Paris talks were to be met, much of the permafrost would melt and release its carbon into the atmosphere. Maybe the time is ripe for visionaries who can come up with groundbreaking ideas to solve the problems facing us. On the other hand, such visionaries may just be unworldly dreamers standing in the way of other solutions.

I stand on the black slope at Duvanny Yar, up to my knees in slurry, slipping and sliding on the frozen earth beneath. While we're there, Nikita finds part of a mammoth's tooth, the same size as a carton of milk. A whole tooth can weigh nearly two kilos. I examine it from different angles, while at the same time attempting to keep the nastiest mosquitoes off my hands. It is at least 14,000 years old, though its age is probably nearer to 20,000 or 30,000.

One of the most fascinating aspects of this earth bank, with its sluggish rivulets of viscous mud, is that any spot on the slope that offers even minimal stability — even if only for a short time — is soon covered in green. These patches form almost absurdly showy islands in the oozing mud. The soil is rich and harbours seeds that have been frozen for as long as the mammoth's tooth. The oldest frozen seed from which scientists have managed to grow a plant comes from this very spot. Virtually all the grass and flowers I can see have grown from seeds thousands of years old, which have germinated as this particular bank of earth thaws out. For a short period, the ground is covered by grass and flowers blooming in the midst of the slurry. When the river changes direction a few years hence and stops cutting into the ground at this spot, the slope will end up covered in moss, willow, and larch, just like the rest of the surrounding landscape — unless the Zimovs have managed to expand their park by then.

'This place is like a zombie; it's where ancient nature returns to life for a while,' says Nikita.

CONCLUSION
Life Will Find a Way

There is no way in which a lost species can really be brought back to life. The nearest thing we can manage is a substitute.

With species like the northern white rhino, the substitute will be genetically identical. What will have been lost is of a different order. We will have lost what might be termed the culture of that species — the things rhino calves learn from their parents and the herd they belong to. But a resurrected rhinoceros released into the wild will be as 'natural' as an animal of any other species raised in a zoo and subsequently released. This has been done many times, with both good and bad outcomes. There's little to suggest that this case would be any different, just because the cells from which the animals were generated had spent a few years in a freezer. There's nothing to prevent new rhino calves from acquiring knowledge and passing it on to their young in turn. This may be the beginning of a new rhinoceros culture.

With species such as mammoths, passenger pigeons, and aurochs, there will be a much wider gap between substitute and original. The question is: how much does that matter? If Siberia becomes home to a large, shaggy, tusked creature with a trunk, which lives in herds and knocks over trees, will that creature be a mammoth? Will it be a mammoth because it fulfils the role of a mammoth, and because we will think of the word 'mammoth' on seeing it? Or will it fail to qualify as a mammoth because it will not be a descendant of the creatures that roamed these parts 10,000 years ago?

At the research station in Siberia, Sergey Zimov and I stand gazing out over the landscape. The station stands on a rise, and you can look out a long way over the meandering river, the sparse forests of larches, and the marshes, where the ground is too wet to support trees. We can see no houses, boats, or roads. I think this is one of the most beautiful places I have ever seen, but Sergey, standing next to me, heaves a sigh.

'People have to understand that this isn't nature. It's a cemetery, a rubbish tip. It's the pathetic remains of an ecosystem that disappeared 14,000 years ago.'

A great many of the researchers I talk to say more or less the same thing. We live in an impoverished world, stripped of much of its biodiversity.

'I used to think de-extinction was about bringing back the past, but now I see it as a way for nature to recover what we owe it in extinct species,' said Ben Novak in Santa Cruz, when I quizzed him about his main reasons for resurrecting the passenger pigeon.

The idea of being able to recover those lost riches is incredibly appealing. But this is where the big problem lies. Would we feel differently about a resurrected species, and would it change our view of nature? Some of our most ancient narratives as humans revolve around the struggle between civilisation and nature, around humanity's efforts to tame the wild and establish a place for culture. Perhaps we have now reached that stage. We have made ourselves masters of nature and are forever moulding it in a myriad ways.

'Do we want to keep mini-rhinos as pets? Do we want to have pet cats that look like tigers? If we have tiger-cats, will we still want to have full-sized tigers living in the wild? What do we want the world to look like?' When I met Oliver Ryder, he put these rhetorical questions to me to show the power we have — for better or for worse — over the natural world. He didn't want to answer his own questions; in his view, it's the responsibility of society as a whole to take the appropriate decisions. He hopes we will become a species that actively enhances global biodiversity; he wants us to make the world blossom rather than deplete it inexorably, as we have done so far.

There is one aspect common to nearly all de-extinction projects: the animals or plants they aim to recreate are charismatic species that excite and fascinate people — mammoths, passenger pigeons, dinosaurs, and majestic chestnut trees. That is hardly a coincidence, since the aim is to attract people's interest and spark off discussion of the technology involved. However, most of the world's research is conducted on mice, fruit flies, and thale cress: small, humdrum species that are easy to manage, but which seldom inspire widespread affection. Scientifically speaking, a resurrected Ice Age rat would be no less impressive than a mammoth, but it would produce considerably fewer headlines.

'The mammoth is a charismatic creature, and people love it. Children are always writing to me — they're excited by this particular project,' says George Church when I ask why he wants to revive the mammoth, rather than any other animal.

So here I am now, my thoughts in a state of flux; I still can't make my mind up. Is trying to resurrect extinct species a good idea? There are sound arguments on both sides. Where you, the reader, position yourself probably depends more on your feelings — as psychologist Susan Clayton argues — than on objective arguments and reason. However, we can try to weigh up the various pros and cons.

There are plenty of arguments that it would be a very bad idea to recreate extinct creatures. The following, in my view, are the three strongest ones.

1. Releasing species into nature involves major risks, and there's no shortage of examples of invasive species that have created huge problems. It is difficult for today's scientists to foresee the consequences of releasing animal species, even those that have been studied in detail. That problem will be far more acute in the case of new species about which we know far less. This is a real scientific risk that has very little to do with our feelings.

2. The relationship between humankind and nature will change. This is what Susan Clayton was talking about when she said that our perception of nature could be altered. Whether that will be significant or not is hard to say. If I were walking through one of the lovely beech forests in southern Sweden, and I looked up through the foliage and spotted a bird whose grandparents had been generated in a lab, would that make me view it any differently? Would I think the forest it was flying through was any less majestic or stately? Or would I feel thankful that the bird still existed? There is also the risk of encouraging people to be careless. If we start thinking it's a trivial matter to resurrect extinct species, we may stop taking action to save the animals and plants under threat today.

3. There is a risk of being dazzled by the opportunities, the tools, and the new technologies, without really thinking about how to put them to best use, and about when we should perhaps refrain from using them. 'If all you have is a hammer, everything looks like a nail,' as Jeanne Loring said when I was talking to her about rhinos and the risks she thought de-extinction might entail. It is human hubris and blindness that can cause problems — the classic fable of Prometheus.

There are also some powerful arguments for resurrecting lost species. In my view, there are three excellent reasons to do so:

1. The new species may potentially enrich the ecosystems of which they are a part. These new creatures could affect their ecosystems in such a way as to benefit many other species. They could act as catalysts, quite simply helping to make their ecosystems healthier. It is this hope that motivates Ben Novak to focus on passenger pigeons rather than on some other bird that would have less of an impact on forests.

2. These projects give people hope and draw attention to possible ways of making the world a better place. Phil Seddon, the species conservationist, stresses that one of the main problems we face today is the feeling that everything is going to hell in a handcart anyway, so the only thing conservationists can do is postpone the inevitable catastrophe. If the extinct northern white rhino were to rise again from the dead, that feeling could change. Optimism, hope, and belief in the future could inspire fresh interest in saving what we have today. This could provide a striking instance of how humans can use new knowledge to make the world better.

3. We will learn a great deal along the way. Regardless of whether we eventually have mammoths lumbering around in Siberia, or whether miniature dinosaurs go on sale in pet shops, the global sum of knowledge is bound to increase through the process itself. All the projects concerned will depend on scientific breakthroughs, and the resultant knowledge can be used in a range of other areas. Ben Novak drew a comparison with the space race. The research conducted to enable us to send people to the moon produced various other advances. One was to show that it was actually possible for people to travel into space — and that, I think, may actually have been the main achievement. In my eyes, this is the strongest argument for pursuing apparently impossible goals such as de-extinction.

'I think it's unfortunate that people think every single scientific project has to have a human application, that it has to put food on your table or gas in your car. It's a pity, but our society seems to be moving away from the whole idea of working in science just because it's a cool thing to do. There's a shift away from scientists spending all their time discovering all there is to be discovered. That's such a precious thing — finding out so much about the world and the universe we live in,' says Jack Horner, when I try to dig deeper into why he wants to build a dino-chicken.

When I asked my non-scientist friends why they think anyone would want to resurrect the mammoth, the reply was mostly: 'Because they can!' Few of the scientists working in the field draw attention to this aspect, but I think it's significant. These projects are driven by curiosity, passion, and the desire to achieve the impossible. These are wonderful wellsprings of motivation, and it is these scientists' passion that has brought me most joy in the course of my research.

This is a book about the last, and the first. When I began writing about humanity's attempts to revive species, I thought my book would focus on nostalgia and the yearning for a vanished world. I discovered that it has more to do with the future, with the present, in which we humans have made ourselves nature's masters — and with scientists' unbridled desire to discover the new.

Acknowledgements

I have heard it said that it takes a village to raise a child. In my opinion, this applies equally, and perhaps even more, to writing a book. This one would not have been possible without the help of a great many people, and I am incredibly grateful for all the support I have been given.

My brilliant editor, Lisa, together with Emma and Christer at my Swedish publishing house, Fri Tanke, has been unstinting in helping me make this book as good as possible. I am also very grateful to Fri Tanke for believing in my idea, which I sent off on spec just before midnight one very grey evening in November 2014.

I would like to thank all the scientists who gave their time to be interviewed, answer my questions, and help me understand the projects they are working on and their outcomes. No scientist works in isolation; all of them are part of a team or group. Although, in the interests of readability, I have confined myself to mentioning only the key individuals in each project, it goes without saying that all the projects involve many other researchers.

While I was carrying out my research for the book, I was helped by many people who are not mentioned by name in its pages. I would like to say a special thank you to Maria Mostadius, curator of the zoological collections at Lund University. It was marvellous to be able to spend an afternoon in her company and familiarise myself with the aurochs skeleton and the passenger-pigeon specimens at Lund.

Everything has to begin somewhere, and I would also like to thank the two teachers who I think were the main reason I decided I wanted to become a science journalist. Thank you, Ragnar, my wonderful marine-biology teacher in high school. Your curiosity and enthusiasm were incredibly infectious, and they are still with me. Thank you, Karin, for taking care of me when I came to *Dagens Nyheter* as a confused novice. You taught me what is expected of a science journalist, starting with the basics, and you gave me challenges that enabled me to grow and improve so much more than I would ever have done on my own.

I would like to thank all my fantastic, fabulous, weird, and wonderful friends for supporting me throughout this whole process. Thank you for listening to my outpourings on everything from mammoth dung to the Epic of Gilgamesh. Thank you for your patience when I was out of reach for weeks on end, only to turn up later like an annoying ghost, needing someone to give me a hug while I sat crying, in the conviction that the book was never, ever going to materialise. Thank you for all the cups of tea and all the encouraging feedback. Thank you for supporting me through the book's tortuous birth pangs. Thank you for reading through successive drafts and making intelligent and helpful comments.

It goes without saying that I want to say an extra thank you for the website mammutkvinnan.se ('The Mammoth Woman'), which made me a professional mammoth poet of sorts. Producing 20 lines of mammoth poetry to order, in iambic pentameters, may well be one of the more unlikely experiences of my life. I can't resist including the first four lines here.

> *I Kolyma flods kalla mörka vatten*
> *Där sjunkna jättar ligger pö om pö*
> *Dit dyker djärva män i kalla natten*
> *De söker skatt som lämnats där att dö.*

An English approximation to this is:

> *Cold and dark the Kolyma flows*
> *O'er sunken giants left there to die.*
> *The bold nocturnal diver knows*
> *He may find treasure by and by.*

I would like to thank my wonderful parents, who were incredibly helpful and supportive while I was writing the book, lending a hand with everything from photography to fact-checking, and helping me structure my work.

Thank you, Tobias; as usual, you have meant more than anyone else. None of this would have been possible without you.

List of Illustrations

Sources, Notes, and Further Reading

Since a large number of my sources are available only on the internet or are much easier to find online than anywhere else, the whole of this section can be read on the book's website (www.kornfeldt.se/mammoth/), which has links to all the references. Much of what I list here is further reading for anyone with an interest in finding out more about the various projects and ideas presented in the book.

INTRODUCTION: *A Whole New World*

1. The myth of Prometheus and fire as the source of civilisation is the subject of many ancient texts. An example is *Prometheus Bound*, which dates back to 400 BC and is traditionally attributed to Aeschylus. Plato also mentions the myth.

2. In 1818, Mary Shelley published *Frankenstein; or, The Modern Prometheus* anonymously in London. Her name did not appear on the novel until the second edition, which came out in France in 1823.

3. The first experiments showing that electricity could make corpses twitch were conducted by Luigi Galvani in 1780. An article about these experiments, with contemporary illustrations of the tests, is 'Animal Electricity, circa 1781' (September 2011) *The Scientist*, https://www.the-scientist.com/?articles.view/articleNo/31078/title/Animal-Electricity--circa-1781/

4. A golem is a creature from Jewish folklore. They are moulded by people, generally out of clay, and are endowed with life of a kind when a sacred word is placed inside their heads. The best-known story tells of a 16th-century rabbi who creates a golem that becomes a servant in the rabbi's house. Everything goes well for a while, but in the end the golem runs amok in the town. The rabbi removes the magic word and swears never again to imitate God by creating life.

5. The film *Jurassic Park* is based on the book of the same name by Michael Crichton, published in 1990. It was not the first book about 'resurrected dinosaurs running amok'; John Brosnan's *Carnosaur* came out in 1984. That was also made into a film, shown for the first time in 1993.

6. The Lazarus Project, whose goal is to resurrect the extinct gastric-brooding frog, also known as the platypus frog (genus: *Rheobatrachus*), is based at the University of New South Wales. The project is led by Michael Archer. More on this project can be found in 'The Lazarus Project: scientists' quest for de-extinction' (April 2015) *The Sydney Morning Herald*, http://www.smh.com.au/technology/sci-tech/the-lazarus-project-scientists-quest-for-deextinction-20150417-1mng6g.html

7. Michael Archer has given a TEDx talk on the Lazarus Project: https://www.ted.com/talks/michael_archer_how_we_ll_resurrect_the_gastric_brooding_frog_the_tasmanian_tiger

8. Eugene Schieffelin was a very intriguing character in many ways and the subject of many interesting articles. An example is: 'The Shakespeare Fanatic Who Introduced All of the Bard's Birds to America' (May 2014) *Pacific Standard*, https://psmag.com/environment/shakespeare-fanatic-introduced-bards-birds-america-82279
    Another is: '100 Years of the Starling' (September 1990) *The New York Times*, http://www.nytimes.com/1990/09/01/opinion/100-years-of-the-starling.html

CHAPTER 1: *Summer in Siberia*

9. The history of Chersky and the whole of eastern Siberia is fascinating, as is the town's development after the fall of the Soviet Union. Here is an Associated Press news article about the time after the collapse of the USSR: 'Isolated Siberian Town Shrivels after Soviet Era' (2011) http://www.foxnews.com/world/2011/01/08/isolated-siberian-town-shrivels-soviet-era.html

10. The research station's homepage is: http://terrychapin.org/station.html

11. Many books have been written about mammoths, their evolution and ecology, and early humans' relationship with them. A good overview is given by Lister & Bahn: *Mammoths: giants of the Ice Age* (2007) Frances Lincoln.

12. Neanderthals used mammoth bones to build houses. Research into the subject is summarised in a news article: 'Neanderthal Home Made of Mammoth Bones Discovered in Ukraine' (December 2011) *Quaternary International*, vol. 247, pp. 1–362, https://phys.org/news/2011-12-neanderthal-home-mammoth-bones-ukraine.html

13. The scientific article in which Nikita and Sergey Zimov try to calculate how many animals lived on the mammoth steppe and compare the result with the population of Africa: 'Mammoth Steppe: a high-productivity phenomenon' (December 2012) *Quaternary Science Reviews*, vol. 57, pp. 26–45, http://www.sciencedirect.com/science/article/pii/S0277379112003939

14. Calculations of when the first humans came to Siberia are based on the carbon 14 dating method: 'The Yana RHS Site: humans in the Arctic before the last glacial maximum' (2004) *Science*, http://science.sciencemag.org/content/303/5654/52

15. New scientific articles about what happened to the mammoths appear regularly. An example is: 'Abrupt Warming Events Drove Late Pleistocene Holarctic Megafaunal Turnover' (July 2015) *Science*, http://science.sciencemag.org/content/349/6248/602

16. Beth Shapiro writes about the extinction of the mammoth in the following scientific article: 'Pattern of Extinction of the Woolly

Mammoth in Beringia' (June 2012) *Nature Communications*, vol. 3, https://www.nature.com/articles/ncomms1881

'Remaining continental mammoths, now concentrated in the north, disappeared in the early Holocene with development of extensive peatlands, wet tundra, birch shrubland and coniferous forest. Long sympatry in Siberia suggests that humans may be best seen as a synergistic cofactor in that extirpation. The extinction of island populations occurred at ~4 ka. Mammoth extinction was not due to a single cause, but followed a long trajectory in concert with changes in climate, habitat and human presence.'

See also: 'Life and Extinction of Megafauna in the Ice-Age Arctic' (September 2015) *Proceedings of the National Academy of Sciences of the United States of America (PNAS)*, vol. 112, http://www.pnas.org/content/112/46/14301.full

17. Questions about whether mammoths lived at the same time as the pyramids were being built proliferate on the internet, and I have seen many different answers. This is mine. The last mammoths died out about 4,000 years ago, and the pyramids at Giza were completed by around 2,560 BC (just over 4,500 years ago). By that time, there were no longer any mammoths living on the mainland, only on remote islands. See: 'Radiocarbon Dating Evidence for Mammoths on Wrangel Island, Arctic Ocean' (1995) *Radiocarbon*, vol. 37, pp. 1–6, https://journals.uair.arizona.edu/index.php/radiocarbon/article/viewFile/1640/1644

18. The trade in mammoth tusks is both open and covert. The estimate of 55 tonnes comes from a news article in *National Geographic*, which is well worth reading and beautifully illustrated: 'Of Mammoths and Men' (April 2013), http://ngm.nationalgeographic.com/2013/04/125-mammoth-tusks/larmer-text

'Nearly 90 percent of all mammoth tusks hauled out of Siberia — estimated at more than 60 tons a year, though the actual figure may be higher — end up in China, where legions of the newly rich are entranced by ivory.'

19. Beth Shapiro has written a book about the possibility of resurrecting a mammoth, in which she describes exactly how scientists go about piecing together ancient DNA: *How to Clone a Mammoth: the science of de-extinction* (2015) Princeton University Press.

20. Sequencing the mammoth genome at the Swedish Museum of Natural History: 'Complete Genomes Reveal Signatures of Demographic and Genetic Declines in the Woolly Mammoth' (May 2015) *Current Biology*, vol. 25, pp. 1395–1400.

CHAPTER 2: *Who Wants to Build a Mammoth?*

21. The Permafrost Kingdom's website is not very good. However, the Yakutia region's tourism site contains a photo gallery with pictures of the frozen caves: http://www.yakutiatravel.com/photo-galery/permafrost-kingdom

22. The scientific article describing the mammoth head in Yakutsk and how it was found is: 'The Yukagir Mammoth: brief history, 14c dates, individual age, gender, size, physical and environmental conditions, and storage' (2006) *Scientific Annals, School of Geology, Aristotle University of Thessaloniki*, vol. 98, pp. 299–314, http://geolib.geo.auth.gr/index.php/sasg/article/view/7524/7281

23. Here are three articles about the three mammoth calves, including pictures of them:
Lyuba: 'Ice Baby' (May 2009) *National Geographic*, http://ngm.national geographic.com/2009/05/mammoths/mueller-text
Zhenya: '"Zhenya" Mammoth Find in North Russia, Biggest in 100 Years, Made By 11-Year-Old Evgeny Salinder' (October 2012) *Huffington Post*, http://www.huffingtonpost.com/2012/10/04/zhenya-mammoth-find-russia_n_1940791.html
Dima: 'Woolly Mammoth: secrets from the ice' (April 2012) BBC, http://www.bbc.co.uk/nature/17525074

24. Dolly, the cloned sheep, was born on 5 July 1996. The researchers in charge of the experiment were Ian Wilmut and Keith Campbell, from the Roslin Institute at the University of Edinburgh. The scientific

article on the experiment is: 'Viable Offspring Derived from Foetal and Adult Mammalian Cells' (February 1997) *Nature*, vol. 385, pp. 810–13, https://www.nature.com/articles/385810a0

25.   *Nature* on the scandal surrounding Hwang Woo-suk: http://www. nature.com/news/specials/woo-suk-hwang-revisited-1.14521
See also: 'Disgraced Korean Scientist Hwang Woo-suk Loses Legal Battle over Mammoth Cloning Tech' (15 August 2017) *The Korea Herald*, http://www.koreaherald.com/view. php?ud=20170815000181

26.   An example of criticism of attempts to find living cells: 'Cloning a Woolly Mammoth: good science or vanity project?' (14 March 2012) *Slate*, http://www.slate.com/blogs/future_tense/2012/03/14/ cloning_a_woolly_mammoth_hwang_woo_suk_and_other_ scientists_attempt_to_revive_exinct_species_.html

27.   Akira Iritani's claim that he would clone a mammoth by 2016 may be found here: 'The first mammoth cloning experiment is officially underway' (January 2011) *Gizmodo*, https://io9.gizmodo. com/5735293/the-first-mammoth-cloning-experiment-is-officially-underway
See also: 'Mammoth "Could Be Reborn in Four Years"' (January 2011) *The Telegraph*, http://www.telegraph.co.uk/news/ science/science-news/8257223/Mammoth-could-be-reborn-in-four-years.html

28.   On the possibility of finding living cells, see also: 'Will We Ever Clone a Mammoth?' (June 2012) BBC, http://www.bbc.com/future/ story/20120601-will-we-ever-clone-a-mammoth

29.   See the homepage of George Church's laboratory at: http://arep.med. harvard.edu/gmc/

30.   A great deal has been written about the CRISPR/Cas9 technique. Here is a good explanation of what it may mean for various species: 'Welcome to the CRISPR Zoo' (9 March 2016) *Nature*, https://www. nature.com/news/welcome-to-the-crispr-zoo-1.19537

31. The Chinese study that was the first to show genetic engineering in human embryos: 'CRISPR/Cas9-mediated Gene Editing in Human Tripronuclear Zygotes' (May 2015) *Protein & Cell*, vol. 6, pp. 363–72, http://link.springer.com/article/10.1007%2Fs13238-015-0153-5

32. George Church has not yet published any scientific study proving success in splicing mammoth genes into elephant DNA. However, he has talked about the experiment in a number of interviews, with me and other people. On the basis of his previous work, I choose to believe his claims. 'Mammoth Genomes Provide Recipe for Creating Arctic Elephants' (May 2015) *Nature*, http://www.nature.com/news/mammoth-genomes-provide-recipe-for-creating-arctic-elephants-1.17462

33. There have been a number of studies into the various effects that mammoth genes could have, and experiments have been conducted with mammoth haemoglobin, which functions at low temperatures: 'Substitutions in Woolly Mammoth Haemoglobin Confer Biochemical Properties Adaptive for Cold Tolerance' (May 2010) *Nature Genetics*, vol. 42, pp. 536–40, http://www.nature.com/ng/journal/v42/n6/full/ng.574.html

34. 'Nuclear Gene Indicates Coat-colour Polymorphism in Mammoths' (July 2006) *Science*, vol. 313, p. 62, http:/science.sciencemag.org/content/313/5783/62

35. 'Elephantid Genomes Reveal the Molecular Bases of Woolly Mammoth Adaptations to the Arctic' (July 2015) *Cell Reports*, vol. 12, pp. 217–28, http://www.cell.com/cell-reports/abstract/S2211-1247(15)00639-7

36. Pluripotent stem cells were induced for the first time by Shinya Yamanaka in 2006, a breakthrough for which he was awarded the 2012 Nobel Prize in medicine: 'Induction of Pluripotent Stem Cells from Mouse Embryonic and Adult Fibroblast Cultures by Defined Factors' (August 2006) *Cell*, vol. 126, pp. 663–76, http://www.cell.com/abstract/S0092-8674(06)00976-7

37.   International Union for Conservation of Nature (IUCN) on Asian elephants: http://www.iucnredlist.org/details/7140/0

CHAPTER 3: *Zombie Spring*

38.   On the virus that lay frozen in Siberian ice for 30,000 years (one of many studies of ancient viruses): 'Thirty-thousand-year-old Distant Relative of Giant Icosahedral DNA Viruses with a Pandoravirus Morphology' (March 2014) *PNAS*, vol. 111, pp. 4274–9, https://www.ncbi.nlm.nih.gov/pmc/articles/PMC3964051/

39.   On frozen plant cells, also 30,000 years old and found in Siberia: 'Regeneration of Whole Fertile Plants from 30,000-y-old Fruit Tissue Buried in Siberian Permafrost' (March 2012) *PNAS*, vol. 109, pp. 4008–13, https://www.ncbi.nlm.nih.gov/pmc/articles/PMC3309767/

40.   Water bears (tardigrades or tardigrada) can survive almost anything; they are incredibly fascinating organisms. For an overview, see the *Encyclopedia of Life*: http://eol.org/pages/3204/overview

41.   A portrait of Steward Brand that covers some of the many things he has done in the course of his life: 'Stewart Brand's Whole Earth Catalog, the Book That Changed the World' (May 2013) *The Guardian*, https://www.theguardian.com/books/2013/may/05/stewart-brand-whole-earth-catalog

42.   The homepage of The Long Now Foundation: http://longnow.org

43.   Revive & Restore's homepage includes information about the various scientific conferences the organisation has held and the projects it supports: http://reviverestore.org

44.   Stewart Brand himself has written and spoken a great deal about the need to revive extinct animals. His TED talk, 'The Dawn of de-extinction. Are You Ready?' (February 2013), is available at: https://www.ted.com/talks/stewart_brand_the_dawn_of_de_extinction_are_you_ready
      See also: 'Rethinking Extinction' (April 2015) *Aeon*, https://aeon.co/essays/we-are-not-edging-up-to-a-mass-extinction

45.   A website displaying pictures of animals that have died out over the past 100 years: https://www.davidwolfe.com/animals-extinct-in-100-years/

46.   The IUCN's list of threatened and extinct animal species: http://www.iucnredlist.org

47.   Although there's no consensus on the number of species on earth, one of the most recent major studies suggests that there are approximately 8.7 million species, plus or minus 1.3 million: 'How Many Species Are There on Earth and in the Ocean?' (August 2011) *PLOS Biology*, http://journals.plos.org/plosbiology/article?id=10.1371/journal.pbio.1001127

48.   A good overview of the five major mass extinction events is: 'Big Five Mass Extinction Events' (October 2014) BBC, http://www.bbc.co.uk/nature/extinction_events

49.   For its coverage of previous mass extinction events and a discussion of whether humans are responsible for a sixth one, I would recommend Elizabeth Kolbert's book *The Sixth Extinction: an unnatural history* (2014), Henry Holt & Co.

50.   Estimate of the number of extinctions for which humanity is responsible: 'Global Effects of Land Use on Local Terrestrial Biodiversity' (April 2015) *Nature*, vol. 520, pp. 45–50, https://www.nature.com/articles/nature14324

51.   There are a few studies that suggest the amount of land given over to farming will dwindle in future. An example is: 'Peak Farmland and the Prospect for Land Sparing' (February 2013) *Population and Development Review*, vol. 38, pp. 221–42, https://phe.rockefeller.edu/docs/PDR.SUPP%20Final%20Paper.pdf
      Another is: 'The Effects of Agricultural Technological Progress on Deforestation: what do we really know?' (June 2014) *Applied Economic Perspectives and Policy*, vol. 36, pp. 211–37, https://academic.oup.com/aepp/article-abstract/36/2/211/8311

52.   Other studies warn that cultivable land will reach its limits, with disastrous consequences. An example is: 'Soil Security: solving the global

soil crisis' (October 2013) *Global Policy*, vol. 4, pp. 434–41, http://
onlinelibrary.wiley.com/doi/10.1111/1758-5899.12096/abstract

53.   On the return of forests to Europe: 'Returning Forests Analysed with
the Forest Identity' (September 2006) *PNAS*, vol. 103, pp. 17574–9,
http://www.pnas.org/content/103/46/17574.full
A news article on the situation in the United States: 'New England
Sees a Return of Forests, Wildlife' (August 2013) *Boston Globe*,
https://www.bostonglobe.com/metro/2013/08/31/new-england-sees-
return-forests-and-wildlife/lJRxacvGcHeQDmtZt09WvN/story.html

CHAPTER 4: *A Winged Storm*

54.   There are many books and articles about Martha and her life and
death, such as: 'The Passenger Pigeon', https://www.si.edu/spotlight/
passenger-pigeon
See also: '100 Years after Her Death, Martha, the Last Passenger
Pigeon, Still Resonates' (September 2014) *Smithsonian
Magazine*, https://www.smithsonianmag.com/smithsonian-
institution/100-years-after-death-martha-last-passenger-pigeon-still-
resonates-180952445/

55.   A great deal has also been written about passenger pigeons in general.
Those interested may want to read Joel Greenberg's *A Feathered River
Across the Sky* (2014) Bloomsbury.

56.   Sixty million birds in Sweden: see Richard Ottvall et al., *Fåglarna
i Sverige: antal och förekomst* (2012), Swedish Ornithological
Association.

57.   Much has also been written about the dodo (*Raphus cucullatus*). An
article touching on the fact that many European scientists believed
the bird had never actually existed is: 'Dead as a Dodo: the fortuitous
rise to fame of an extinction icon' (September 2008) *Historical
Biology*, vol. 20, pp. 149–63, http://dodobooks.com/wp-content/
uploads/2012/01/TurveyCheke-2008-Dead-as-a-dodo.pdf

58. Georges Cuvier was an interesting scientist, who gave a public reading of his original work on mammoth fossils, *Mémoires sur les espèces d'éléphants vivants et fossiles*, in 1796. It was published in 1800.

59. Ben Novak's project is mentioned on this website: https://pgl.soe.ucsc.edu

60. Ben's TEDx talk is available at: https://www.youtube.com/watch?v=rUoSjgZCXhe

61. The type of cells that Ben seeks to modify, known as primordial germ cells, have been much studied in birds. One of the most recent breakthroughs in this field was made in the Roslin Institute at the University of Edinburgh, the same research institution that cloned Dolly the sheep. See: 'Cryptopreservation of Specialised Chicken Lines Using Cultured Primordial Germ Cells' (August 2016) *Poultry Science*, vol. 95, pp. 1905–11, https://academic.oup.com/ps/article/95/8/1905/2563774

62. The first study of gene editing in this type of cells was: 'Germline Gene Editing in Chickens by Efficient CRISPR-mediated Homologous Recombination in Primordial Germ Cells' (April 2016) *PLOS One*, http://journals.plos.org/plosone/article?id=10.1371/journal.pone.0154303

63. The first genetically modified plant, a tobacco plant resistant to antibiotics, is described in: 'Expression of Bacterial Genes in Plant Cells' (August 1983) *PNAS*, vol. 80, pp. 4803–7, http://www.pnas.org/content/80/15/4803.full.pdf

64. Genetically modified bacteria are used to produce medicines such as insulin: 'Protein Therapeutics: a summary and pharmacological classification' (January 2008) *Nature Reviews Drug Discovery*, vol. 7, pp. 21–39, https://www.nature.com/articles/nrd2399; 'Therapeutic Insulins and Their Large-scale Manufacture' (December 2004) *Applied Microbiology and Biotechnology*, vol. 67, pp. 151–9, http://link.springer.com/article/10.1007%2Fs00253-004-1809-x

65. On insects that develop resistance to Bt crops: 'Insect Resistance to Bt Crops: lessons from the first billion acres' (June 2013) *Nature*

Biotechnology, vol. 31, pp. 510–23, https://www.nature.com/articles/nbt.2597

66. Study summarising the information available on the impact on human health of eating GM crops: 'Published GMO Studies Find No Evidence of Harm when Corrected for Multiple Comparisons' (January 2016) *Critical Reviews in Biotechnology*, vol. 37, pp. 213–17, http://www.tandfonline.com/doi/pdf/10.3109/07388551.2015.1130684

67. On the expiry of patents on various GM crops: 'As Patents Expire Farmers Plant Generic GMOs' (July 2015) *MIT Technology Review*, https://www.technologyreview.com/s/539746/as-patents-expire-farmers-plant-generic-gmos/

68. The first scientific article on golden rice was published in 2000: 'Engineering the Provitamin A (β-carotene) Biosynthetic Pathway into (Carotenoid-Free) Rice Endosperm' (January 2000) *Science*, vol. 287, pp. 303–5, http://science.sciencemag.org/content/287/5451/303

CHAPTER 5: *New Kid on the Block*

69. The scientific article on the cloning of Celia: 'First Birth of an Animal from an Extinct Subspecies (*Capra pyrenaica pyrenaica*) by Cloning' (April 2009) *Theriogenology*, vol. 71, pp. 1026–34, https://www.ncbi.nlm.nih.gov/pubmed/19167744

70. Alberto's TEDx talk on the bucardo and the experiments conducted by his team: 'The First De-extinction' (April 2013), https://www.youtube.com/watch?v=5eMqEQw9Fbs

71. Dolly the sheep was named after Dolly Parton.

72. Several articles have been published that are critical of the attempts to resurrect the bucardo, such as: 'The Arguments against Cloning the Pyrenean Wild Goat' (November 2014) *Conservation Biology*, vol. 28, pp. 1445–6, http://onlinelibrary.wiley.com/doi/10.1111/cobi.12396/abstract

CHAPTER 6: *The Rhino That Came in from the Cold*

73. The 'white' part of the name 'Northern white rhinoceros'
     (*Ceratotherium simum*) is problematic. The two subspecies of
     the white rhinoceros are known as the northern white rhino and
     the southern white rhino respectively. The most widely accepted
     explanation for the animal's English name is that 'white' reflects a
     misinterpretation by the British of its description in Afrikaans, *wyd
     muil*, which actually refers to its wide mouth or blunt muzzle. If
     you want to know more about the rhinoceros's name, I recommend
     Kees Rookmaaker's 'Why the Name of the White Rhinoceros Is Not
     Appropriate' (2003) *Pachyderm*, vol. 34, pp. 88–93, http://www.
     rhinoresourcecenter.com/pdf_files/117/1175858144.pdf

74. On poaching and the high prices for rhinoceros horn: 'Which is Most
     Valuable: Gold, Cocaine, Or Rhino Horn?' (May 2015) *I Fucking
     Love Science*, http://www.iflscience.com/plants-and-animals/which-
     most-valuable-gold-cocaine-or-rhino-horn/

75. The homepage of the San Diego Frozen Zoo: http://institute.
     sandiegozoo.org/resources/frozen-zoo%C2%AE

76. Oliver Ryder's text on resurrecting extinct species, 'Designing the
     Destiny of Biological Diversity' (2013) is available at: https://www.
     humansandnature.org/conservation-extinction-oliver-ryder

77. For Oliver Ryder's TEDx talk, 'Genetic Rescue and Biodiversity
     Banking' (April 2013), see: https://www.youtube.com/
     watch?v=EVzhs1WjzGg

78. On stem cells, see the references for Chapter 2.

79. Jeanne Loring's laboratory, the Centre for Regenerative Medicine, has
     its homepage at: http://www.scripps.edu/loring/

80. Here is the scientific article on how the team succeeded in generating
     stem cells from the white rhino: 'Generation of Induced Pluripotent
     Stem Cells from Mammalian Endangered Species' (2015) *Cell
     Programming*, vol. 1330, pp. 101–9, https://link.springer.com/
     protocol/10.1007/978-1-4939-2848-4_10

81. The Chinese study that claims success in generating early mouse sperm is: 'Complete Meiosis from Embryonic Stem Cell-derived Germ Cells In Vitro' (March 2016) *Cell Stem Cell*, vol. 18, pp. 330–40, http://www.cell.com/cell-stem-cell/abstract/S1934-5909(16)00018-7 For a critique of the study, see: 'Researchers Claim to Have Made Artificial Mouse Sperm in a Dish' (2016) *Nature*, http://www.nature.com/news/researchers-claim-to-have-made-artificial-mouse-sperm-in-a-dish-1.19453

82. Experiments in which scientists succeeded in inseminating an endangered species of animal with sperm that had been frozen for 20 years are documented in: 'Recovery of Gene Diversity Using Long-term Cryopreserved Spermatozoa and Artificial Insemination in the Endangered Black-footed Ferret' (August 2015), *Animal Conservation*, vol. 9, pp. 102–11, http://onlinelibrary.wiley.com/doi/10.1111/acv.12229/abstract

83. Summary of the possibility of generating human ova and sperm from stem cells: 'Human Primordial Germ Cells in a Dish' (January 2015) *Nature Reviews Molecular Cell Biology*, vol. 16, https://www.nature.com/articles/nrm3945

84. The most recent scientific article by Jeanne Loring and Oliver Ryder that summarises their plans for the white rhino in the future can be found at: 'Rewinding the Process of Mammalian Extinction' (May 2016) *Zoo Biology*, http://onlinelibrary.wiley.com/doi/10.1002/zoo.21284/abstract

85. The company that's planning to sell genetically modified pigs and koi carp is called BGI. 'Gene-edited "Micropigs" to Be Sold as Pets at Chinese Institute' (September 2015) *Nature*, http://www.nature.com/news/gene-edited-micropigs-to-be-sold-as-pets-at-chinese-institute-1.18448

86. Summary of CRISPR for the modification of pets: 'Welcome to the CRISPR Zoo', https://www.nature.com/news/welcome-to-the-crispr-zoo-1.19537

CHAPTER 7: *'It's Not Quite That Simple'*

87. Phil Seddon's homepage is at: http://www.otago.ac.nz/zoology/staff/
     otago008934

88. Phil Seddon's scientific article on the selection of species for
     revival: 'Reintroducing Resurrected Species: selecting de-extinction
     candidates' (March 2014) *Trends in Ecology & Evolution*, vol. 29, pp.
     140–7, http://www.cell.com/trends/ecology-evolution/fulltext/
     S0169-5347%2814%2900021-4

89. 'Genetic Rescue to the Rescue' (November 2014) *Trends in Ecology
     & Evolution*, vol. 30, pp. 42–49, https://www.researchgate.net/
     publication/268821953_Genetic_rescue_to_the_rescue

90. On giant tortoises in the Indian Ocean: 'Assessing the Potential
     to Restore Historic Grazing Ecosystems with Tortoise Ecological
     Replacements' (June 2013) *Conservation Biology*, vol. 27, pp. 690–
     700, http://onlinelibrary.wiley.com/doi/10.1111/cobi.12087/abstract

91. Interview in *Wired* with Maura O'Connor: 'Biologists Could Soon
     Resurrect Extinct Species. But Should They?' (September 2015)
     Wired, https://www.wired.com/2015/11/biologists-could-soon-
     resurrect-extinct-species-but-should-they/

92. Maura O'Connor's book is entitled *Resurrection Science: conservation,
     de-extinction, and the precarious future of wild things* (2015) St Martin's
     Press.

93. Interesting report on cat experiments in New Orleans: 'Where Cats
     Glow Green: weird feline science in New Orleans' (November 2013)
     The Verge, https://www.theverge.com/2013/11/6/4841714/where-
     cats-glow-green-weird-feline-science-acres-in-new-orleans

94. A scientific article on the cloned wildcat: 'Birth of African Wildcat
     Cloned Kittens Born from Domestic Cats' (October 2004) *Cloning
     Stem Cells*, vol. 6, pp. 247–58, https://www.ncbi.nlm.nih.gov/
     pubmed/15671671

95. There have been a number of attempts to clone endangered species of
     animals.

The banteng: 'Collaborative Effort Yields Endangered Species Clone' (April 2003) *Advanced Cell Technology*, https://www.prnewswire. com/news-releases/collaborative-effort-yields-endangered-species-clone-70813392.html

The gaur: 'Cloning of an Endangered Species (*Bos gaurus*) Using Interspecies Nuclear Transfer' (2000) *Cloning*, vol. 2, pp. 79–90, http://media.longnow.org/files/2/REVIVE/Cloning%20of%20 an%20Endangered%20Species.pdf

The mouflon: 'Genetic Rescue of an Endangered Mammal by Cross-species Nuclear Transfer Using Post-mortem Somatic Cells' (October 2001) *Nature Biotechnology*, vol. 19, pp. 962–4, https:// www.ncbi.nlm.nih.gov/pubmed/11581663

96.   For a summary of the state of the art of interspecies animal cloning, see: 'Interspecies Somatic Cell Nuclear Transfer: advancements and problems' (October 2013), *Cell Reprogramming*, vol. 15, pp. 374–84, https://www.ncbi.nlm.nih.gov/pubmed/24033141

97.   A good popular science article on whether cloning can help save endangered animals is: 'Will Cloning Ever Save Endangered Animals?' (March 2013) *Scientific American*, https://www.scientificamerican. com/article/cloning-endangered-animals/

CHAPTER 8: *God's Toolkit*

98.   The homepage for coral spawning in Australia is: https://www.aims. gov.au/seasim-coral-spawning-activities

99.   Madeleine van Oppen's homepage: http://data.aims. gov.au/staffcv/jsf/external/view.xhtml;jsessionid= 447C951DF26A2B6D5ACA6FA8444FFE5C?partyId=100000442

100. Layla and leukaemia: 'Leukaemia Success Heralds Wave of Gene-editing Therapies' (November 2015) *Nature News*, http://www. nature.com/news/leukaemia-success-heralds-wave-ofgen-editing-therapies-1.18737

101. Summary of possible ways in which gene editing could be used for conservation purposes: Thomas et al.: 'Gene tweaking for

conservation' (September 2013) *Nature*, vol. 501, pp. 485–486, http://www.nature.com/news/ecology-gene-tweaking-for-conservation-1.1379002. Summary of the conference of geneticists and conservation biologists organised by the Long Now (April 2015): Case studies, http://reviverestore.org/case-studies/

102. Report on gene drives: 'Nu kan vi styra över domedagsgenen' (November 2015) *Forskning & Framsteg*, http://fof.se/tidning/2015/10/artikel/nu-kan-vi-styra-over-domedagsgenen

103. Attempts to provide protection against gene drives that have been disseminated by mistake: 'Safeguarding CRISPR-Cas9 Gene Drives in Yeast' (November 2015) *Nature Biotechology*, vol. 33, pp. 1250–5, https://www.nature.com/articles/nbt.3412

104. An interesting article about the ways in which synthetic biology can be used to solve ecological problems, and the risks that that entails: 'Synthetic Biology and Conservation of Nature: wicked problems and wicked solutions' (April 2013) *PLOS Biology*, http://journals.plos.org/plosbiology/article?id=10.1371/journal.pbio.1001530

CHAPTER 9: *The Growing Dead*

105. The homepage of the American Chestnut Research and Restoration Project is: http://www.esf.edu/chestnut/

106. William Powell's homepage is: http://www.esf.edu/EFB/powell/

107. William's TEDx talk is: 'Reviving the American Forest with the American Chestnut' (April 2013), https://www.youtube.com/watch?v=WYHQDLCmgyg

108. An article by William Powell about the chestnut is: 'The American Chestnut's Genetic Rebirth' (March 2014), *Scientific American*, https://www.scientificamerican.com/article/the-american-chestnut-genetic-rebirth/

109. A scientific article about the genetically modified chestnuts: 'Improving Rooting and Shoot Tip Survival of Micropropagated Transgenic American Chestnut Shoots' (February 2016) *HortScience*, vol. 51, pp. 171–6, http://hortsci.ashspublications.org/content/51/2/171.short

110. A report on re-establishing an entire forest: 'Resurrecting a Forest' (November 2013) *The Loom*, http://phenomena.nationalgeographic. com/2013/03/11/resurrecting-a-forest/
Progress report: 'Moving Closer to 10,000 Trees' (October 2017), http://www.esf.edu/chestnut/documents/10000-chestnut-challenge-report-2017.pdf

111. An article by Johanna Witzell about using non-harmful fungi as a means of protection against plant diseases: 'Ecological Aspects of Endophyte-based Biocontrol of Forest Diseases' (October 2013) *Advances in Endophytic Research*, pp. 321–333, https://link.springer. com/chapter/10.1007/978-81-322-1575-2_17

112. A scientific article about genetic protection against ash dieback: 'Genetic Resistance to *Hymenoscyphus pseudoalbidus* Limits Fungal Growth and Symptom Occurrence in *Fraxinus excelsior*' (July 2011) *Forest Pathology*, vol. 42, pp. 69–74, http://onlinelibrary.wiley.com/ doi/10.1111/j.1439-0329.2011.00725.x/abstract

113. The disease now threatening beeches is known as *Phytophthora*. For a summary of its impact on European forests, read: 'Recent Developments in *Phytophthora* Diseases of Trees and Natural Ecosystems in Europe' (2006) *Progress in Research on* Phytophthora *Diseases of Forest Trees*, https://www.forestry.gov.uk/pdf/ Phytophthora_Diseases_Chapter01.pdf/$FILE/Phytophthora_ Diseases_Chapter01.pdf

CHAPTER 10: *If It Walks Like a Duck and Quacks Like a Duck — Is It an Aurochs?*

114. The story about how Hermann Göring personally shot his Heck cattle comes from the Dutch researcher Clemens Driessen, who has studied the Heck brothers. He's the author of 'Back-breeding the Aurochs: the Heck brothers, National Socialism, and imagined geographies for nonhuman Lebensraum', a chapter in Giaccaria & Minca: *Hitler's Geographies* (2016) University of Chicago Press.

115. More about the history of the aurochs in Poland can be found
     in: 'History of the Aurochs (*Bos taurus primigenius*) in Poland'
     (April 1995) *Animal Genetic Resources*, vol. 16, pp. 5–12, https://
     www.cambridge.org/core/journals/animal-genetic-resources-
     resources-genetiques-animales-recursos-geneticos-animales/article/
     history-of-the-aurochs-bos-taurus-primigenius-in-poland/73E5642D
     C0324EC98B52B34C57F9AE92

116. An article about the aurochs horn in the Swedish Royal Armoury
     has appeared in the magazine for friends of the Royal Armoury:
     'Uroxehornet, ett eftertraktat Livsrustkammarföremål' (June 2011)
     Livsrustkammarens Vänner Medlemsblad, http://livrustkammaren.se/
     sites/livrustkammaren.se/files/medlemsblad203620juni2020111.pdf

117. A summary of the Heck brothers' experiment, including quotations
     from both brothers about how successful they considered it to have
     been, may be found in: 'History, Morphology, and Ecology of the
     Aurochs (*Bos primigenius*)' (2002), http://members.chello.nl/~t.
     vanvuure/oeros/uk/lutra.pdf

118. The homepage of the True Nature Foundation is at: https://
     truenaturefoundation.org

119. Henri Kerkdijk-Otten's TEDx talk about the project, 'Restoring
     Europe's Wildlife with Aurochs and Others' (April 2013), can be
     viewed at: https://www.youtube.com/watch?v=Obo9odbGOYU

120. An article on the cattle at Chillingham Castle: 'A Viable Herd of
     Genetically Uniform Cattle' (January 2001), *Nature*, vol. 409, https://
     www.nature.com/articles/35053160

121. An article about the quagga: 'The Quagga and Science: what does the
     future hold for this extinct zebra?' (2013), *Perspectives in Biology and
     Medicine*, vol. 56, http://muse.jhu.edu/article/509324

122. As of the start of 2018, a handful of breeding projects are still active —
     calves are being born and breeds are being combined — but no one is
     yet claiming to have produced a 'perfect' new auroch. A very good blog
     for news: http://breedingback.blogspot.com/

CHAPTER 11: *A Wilder Europe*

123. Uffe Gjøl Sørensen's view of the project at Lille Vildmose:
      'Vildokserne ved Lille Vildmose 2003–2010. Status rapport med
      anbefalinger til projektets forvaltning' (2010) UG Sørensen Consult,
      http://naturstyrelsen.dk/media/nst/Attachments/Lille_Vildmose_
      anbefalinger_for_naturpleje1.pdf

124. Report on rewilding: 'Förvilda Europa' (May 2012), *Forskning &
      Framsteg*, http://fof.se/tidning/2012/4/forvilda-europa

125. Dissertation on rewilding of Sweden: Pettersson, *'Återförvilda' Sverige?
      En studie av rewilding som strategi för att bevara kulturlandskapet och
      gynna biologisk mångfald*, thesis for bachelor degree in global studies,
      spring term, 2014, http://nordensark.se/media/1341/examensarbete-
      rewildinghannapettersson.pdf

126. Impact of wild boar on forests: 'Ekologiska och ekonomiska
      konsekvenser av vildsvinens (*Sus scrofa*) återetablering i Sverige' (2013)
      Biology Education Centre, Uppsala University, http://files.webb.
      uu.se/uploader/271/BIOKand-13-025-Duck-Lovisa-Uppsats.pdf

127. There are a number of reports on the European bison released in
      the Carpathian Mountains. *The Guardian* reported on the first ever
      release: 'Return of the European Bison' (May 2014) *The Guardian*,
      https://www.theguardian.com/environment/2014/may/21/-sp-
      european-bison-europe-romania-carpathian-mountains
      A press release from Rewilding Europe when the gates to the
      enclosure were opened and the bison were able to take their first
      steps in freedom: '28 European Bison Now Roaming the Tarcu
      Mountains in the Southern Carpathians' (June 2015), https://www.
      rewildingeurope.com/news/28-european-bison-now-roaming-the-
      tarcu-mountains-in-the-southern-carpathian/
      A press release from Rewilding Europe about their Tauros aurochs
      grazing in Croatia: 'Second Generation of Tauros Now Grazing in
      Lika Plains' (June 2016), https://www.rewildingeurope.com/news/
      second-generation-of-tauros-now-grazing-in-lika-plains/

128. An excellent report on rewilding and the Oostvaardersplassen by Elizabeth Kolbert: 'Recall of the Wild' (December 2012) *The New Yorker*, https://www.newyorker.com/magazine/2012/12/24/recall-of-the-wild

129. A critical report about the problems with the Oostvaardersplassen: 'Holländskt naturexperiment slutade i katastrof' (September 2012) *Svensk Jakt*, https://svenskjakt.se/start/Nyheter/2012/09/hollandskt-naturexperiment-slutade-i-katastrof/

130. Information from Yellowstone National Park about wolves and their impact: 'Wolf Reintroduction Changes Ecosystem' (June 2011), http://www.yellowstonepark.com/wolf-reintroduction-changes-ecosystem/

131. Report on wolves in Denmark: 'Research Project Reveals the Secrets of the Danish Wolf' (June 2015) *Copenhagen Post*, http://cphpost.dk/news/research-project-reveals-the-secrets-of-the-danish-wolf.html

CHAPTER 12: *'Most People Would Call This Totally Insane'*

132. Sergey Zimov's article about the attempts to recreate the mammoth steppe: 'Pleistocene Park: return of the mammoth's ecosystem' (May 2005) *Science*, vol. 308, pp. 796–8, http://www.pleistocenepark.ru/files/Zimov_PleistocenePark_Science.pdf

CHAPTER 13: *A Chicken's Inner Dinosaur*

133. The most ancient hereditary material ever analysed comes from a horse that lived 700,000 years ago: 'World's Oldest Genome Sequenced from 700,000-year-old Horse DNA' (June 2013) *National Geographic*, https://news.nationalgeographic.com/news/2013/06/130626-ancient-dna-oldest-sequenced-horse-paleontology-science/

134. Blood and collagen found in a dinosaur fossil: '75-million-year-old Dinosaur Blood and Collagen Discovered in Fossil Fragments' (June 2015) *The Guardian*, https://www.theguardian.com/science/2015/jun/09/75-million-year-old-dinosaur-blood-and-collagen-discovered-in-fossil-fragments

135. Jack Horner's homepage: http://www.montana.edu/earthsciences/ facstaff/horner.html

136. Jack Horner's TED talk, 'Building a Dinosaur from a Chicken' (March 2011), is available at: https://www.youtube.com/ watch?v=0QVXdEOiCw8

137. Arkhat Abzhanov's homepage is at: http://www.imperial.ac.uk/ people/a.abzhanov

138. Arkhat Abzhanov's scientific article about chickens that develop snouts: 'A Molecular Mechanism for the Origin of a Key Evolutionary Innovation, the Bird Beak and Palate, Revealed by an Integrative Approach to Major Transitions in Vertebrate History' (June 2015) *Evolution*, vol. 69, pp. 1665–77, http://onlinelibrary.wiley.com/ doi/10.1111/evo.12684/abstract

CHAPTER 14: *The Fine Line Between Utopia and Dystopia*

139. Ben Minteer's text about the difficulties with resurrecting species is: 'Is it Right to Reverse Extinction?' (May 2014) *Nature*, vol. 509, http:// www.nature.com/news/is-it-right-to-reverse-extinction-1.15212

140. He has also written a longer piece on the same subject: 'Extinct Species Should Stay Extinct' (December 2014) *Slate*, http://www.slate.com/ articles/technology/future_tense/2014/12/de_extinction_ethics_ why_extinct_species_shouldn_t_be_brought_back.html

141. Susan Clayton's homepage is at: http://discover.wooster.edu/sclayton/

142. A book about the possibility of resurrecting animal species with a focus on ethics: Fletcher, *Mendel's Ark: biotechnology and the future of extinction* (2014) Springer.

143. A book that summarises the ethical issues surrounding the resurrection of extinct species: Oksanen & Siipi (eds), *The Ethics of Animal Re-creation and Modification: reviving, rewilding, restoring* (February 2014) Palgrave.

144. An article about the ethical issues around the recreation of various animals: 'The Ethics of Reviving Long-Extinct Species' (July 2013) *Conservation Biology*, vol. 28, pp. 354–60, http://hettingern.people.

cofc.edu/150_Spring_2015/Sandler_Ethics_of_Reviving_Long_
Extinct_Species.pdf

145. Report on the difficulties with discussing the potential and problems
of a technology that barely even exists yet in practice: 'All This Talk
about De-extinction Is Endangering the Whole Idea' (March 2014)
Motherboard, https://motherboard.vice.com/en_us/article/bmjnvv/
all-this-talk-about-de-extinction-is-endangering-the-whole-idea

CHAPTER 15: *A Melting Giant*

146. Sergey Zimov's scientific article on the amount of carbon in the
permafrost: 'Permafrost and the Global Carbon Budget' (June 2006)
Science, vol. 312, pp. 1612–13, http://imedea.uib-csic.es/master/
cambioglobal/Modulo_V_cod101619/Permafrost%20response.pdf

147. Scientific article about the link between the climate and the carbon
contained in the melting permafrost: 'Climate Change and the
Permafrost Carbon Feedback' (April 2015) *Nature*, vol. 520, pp.
171–9, https://www.nature.com/articles/nature14338

148. An in-depth report, by me, on the melting permafrost in Siberia:
'Svarta hotet' (November 2015) *Forskning & Framsteg*, http://fof.se/
tidning/2015/10/artikel/svarta-hotet

149. The lack of large herbivores causes a great many problems, one of
which is the shortage of dung: 'How Poop Made the World Go
'Round' (November 2015) *The Atlantic*, https://www.theatlantic.com/
science/archive/2015/11/how-the-poop-of-giant-animal-species-kept-
the-world-healthy/413608

*

Below I have collected together some examples of critiques of de-
extinction and the ongoing theoretical debate between various
scientists about the possibility of resurrecting extinct animals. This
makes interesting reading, and I would like to draw attention to a few
articles that may offer a good way into the subject. There are further
discussions of ethics in Chapter 14.

150. Two articles (for and against) and a leader from the magazine *Frontiers of Biogeography*, vol. 6, March 2014: 'De-extinction: raising the dead and a number of important questions'; 'From Dinosaurs to Dodos: who could and should we de-extinct?'; 'De-extinction in a Crisis Discipline'. Link to all: https://escholarship.org/uc/fb/6/1

151. A biological scientist who is extremely positive about the experiments: Greene, 'As Far as We Can Go, as Far as We Want to Go ...', Centre for Humans & Nature, https://www.humansandnature.org/conservation-extinction-harry-w.-greene

152. A biologist who is very critical of the attempts to resurrect animal species: Ehrenfeld, 'Resurrected Mammoths and Dodos? Don't Count on It' (March 2013) *The Guardian*, https://www.theguardian.com/commentisfree/2013/mar/23/de-extinction-efforts-are-waste-of-time-money

153. Another very critical biologist: Ehrlich, 'The Case against De-Extinction: it's a fascinating but dumb idea' (January 2014) *Environment 360*, http://e360.yale.edu/features/the_case_against_de-extinction_its_a_fascinating_but_dumb_idea

154. Yet another very critical biologist: Pimm, 'The Case against Species Revival' (March 2013) *National Geographic*, https://news.nationalgeographic.com/news/2013/03/130312--deextinction-conservation-animals-science-extinction-biodiversity-habitat-environment/